ENERGY SCIENCE, ENGINEERING AND TECHNOLOGY

CLEAN ENERGY SOLUTIONS FROM COAL

ENERGY SCIENCE, ENGINEERING AND TECHNOLOGY

Additional books in this series can be found on Nova's website
under the Series tab.

Additional E-books in this series can be found on Nova's website
under the E-books tab.

ENERGY POLICIES, POLITICS AND PRICES

Additional books in this series can be found on Nova's website
under the Series tab.

Additional E-books in this series can be found on Nova's website
under the E-books tab.

ENERGY SCIENCE, ENGINEERING AND TECHNOLOGY

CLEAN ENERGY SOLUTIONS FROM COAL

MARCUS A. PELT
EDITOR

Nova Science Publishers, Inc.
New York

For permission to use material from this book please contact us:
Telephone 631-231-7269; Fax 631-231-8175
Web Site: http://www.novapublishers.com

Library of Congress Cataloging-in-Publication Data

Clean energy solutions from coal / editor, Marcus A. Pelt.
p. cm.
ISBN 978-1-61324-724-2 (hardcover)
1. Clean coal technologies. 2. Coal--Environmental aspects. 3. Clean energy industries. 4. Green technology. I. Pelt, Marcus A.
TP325.C4875 2011
665.5'384--dc23

2011016762

Published by Nova Science Publishers, Inc. † New York

CONTENTS

PREFACE

The U.S. is currently faced with competing strategic objectives related to energy: energy supply security, economic sustainability and concerns over global climate change. Coal to Liquids (CTL) is a commercial process which converts coal into diesel fuel, producing a concentrated stream of CO2 as a byproduct. Coupling the process with carbon sequestration is relatively inexpensive and results in a fuel with appreciably less life cycle GHG emissions that the average U.S. petroleum-derived diesel. This book explores the CTL and Coal and Biomass to Liquids (CBTL) technologies, which as part of the U.S. energy portfolio, offer a balanced solution to the nation's transportation fuel dilemma, providing affordable fuels from domestic feedstocks, and enabling significant reduction in GHG emissions.

Chapter 1- The United States of America is currently faced with competing strategic objectives related to energy: energy supply security, economic sustainability, and concerns over global climate change. As President-elect Obama alludes, the transportation sector is at the crux of this dilemma: high fuel price volatility directly affects the health of our economy and economic competitiveness, roughly two thirds of our transportation fuels are imported, and transportation is responsible for more carbon dioxide (CO_2) emissions than any other end-use sector of our economy (34% of our total CO_2 emissions by end-use). [1,2]

Chapter 2- Hydrogen has the potential to play a significant role in the nation's energy future, particularly for the production of clean electric power from coal. Production and use of hydrogen in a gasification combined cycle system for stationary power applications will complement the development of next generation hydrogen turbine technology that enables the plant to achieve near-zero pollutant emissions and increased plant efficiency. The Hydrogen from Coal Program's Research, Development, and Demonstration (RD&D) activities include development of hydrogen separation membranes and other advanced technologies that efficiently produce high purity hydrogen for stationary power production. When combined with carbon management technologies such as carbon capture and storage (CCS) and coal-biomass co-utilization, these next generation power plants will achieve significant reductions in greenhouse gas (GHG) emissions with low electricity costs.

Chapter 3- The overall objective of this testing is to develop membrane technologies that achieve target performance of hydrogen separation membranes for use in a gasification system process. NETL needs (1) to know whether developers are approaching/achieving Hydrogen Program Technical Targets, (2) to be able to compare results on an "apples to apples" basis, and (3) to clearly state expectations to contractors. Results may be of value in deciding on down-selections. Another objective is to determine the suitability of each membrane and to assess the compatibility of each membrane's optimum operating conditions for use in plant hardware as it currently exists.

Chapter 4- Coal is a solid fossil fuel with a high carbon content but a low hydrogen content, typically no more than 5–6 percent of the total weight of coal. On a molecular level, it consists of long chains of mostly aromatic hydrocarbon structures. It is mostly associated with the generation of electric power or as a feedstock in the production of steel. However, this versatile, solid rock can be broken down into simple molecules and put back together into many diff erent, useful forms.

In: Clean Energy Solutions from Coal
Editor: Marcus A. Pelt
ISBN: 978-1-61324-724-2

Chapter 1

AFFORDABLE, LOW-CARBON DIESEL FUEL FROM DOMESTIC COAL AND BIOMASS*

National Energy Technology Laboratory

NETL Contact:
Thomas J. Tarka, P.E.
Energy Systems Engineer
Office of Systems, Analyses, and Planning

Prepared by:
Thomas J. Tarka, P.E.
U.S. Department of Energy
National Energy Technology Laboratory
John G. Wimer
U.S. Department of Energy
National Energy Technology Laboratory
Peter C. Balash
U.S. Department of Energy
National Energy Technology Laboratory
Timothy J. Skone, P.E.
U.S. Department of Energy
National Energy Technology Laboratory
Kenneth C. Kern
U.S. Department of Energy
National Energy Technology Laboratory
Maria C. Vargas
U.S. Department of Energy

* This is an edited, reformatted and augmented version of a National Energy Technology Laboratory publication, DOE/NETL-2009/1349, from www.netl.doe.gov, dated January 14, 2009.

National Energy Technology Laboratory
Bryan D. Morreale, Ph.D.
U.S. Department of Energy
National Energy Technology Laboratory
Charles W. White III
Noblis, Inc.
David Gray
Noblis, Inc.

DISCLAIMER

This report was prepared as an account of work sponsored by an agency of the United States Government. Neither the United States Government nor any agency thereof, nor any of their employees, makes any warranty, express or implied, or assumes any legal liability or responsibility for the accuracy, completeness, or usefulness of any information, apparatus, product, or process disclosed, or represents that its use would not infringe privately owned rights. Reference therein to any specific commercial product, process, or service by trade name, trademark, manufacturer, or otherwise does not necessarily constitute or imply its endorsement, recommendation, or favoring by the United States Government or any agency thereof. The views and opinions of authors expressed therein do not necessarily state or reflect those of the United States Government or any agency thereof.

ACKNOWLEDGMENTS

This work would not be possible without the guidance and assistance provided by Daniel Cicero; the extensive process modeling and cost estimation efforts of Charles White and David Gray at Noblis, Inc.; the comprehensive biomass feedstock review and subsequent support from David Ortiz and Henry Willis at the RAND Corporation; guidance from Mark Ackiewicz, Guido B. DeHoratiis, and Lowell Miller in the U.S. DOE's Office of Fossil Energy; and the entirety of the National Energy Technology Laboratory's Office of Systems, Analyses, and Planning, especially Kristin Gerdes and Erik Shuster.

Special mention is also due to Dr. Robert Williams and his team at the Princeton Environmental Institute for their guidance and early analysis and persistent advocacy of the significant advantages of the Coal and Biomass to Liquids (CBTL) process.

EXECUTIVE SUMMARY

> "...this has been our pattern. We go from shock to trance...oil prices go up, gas prices at the pump go up, everybody goes into a flurry of activity. And then the prices go back down and suddenly we act like it's not important, and we start... filling up our SUVs again. And, as a consequence, we never make any progress.
>
> It's part of the addiction, all right...that has to be broken. Now is the time to break it."

President-elect Barack Obama, "60 Minutes" interview, November 16, 2008

The United States of America is currently faced with competing strategic objectives related to energy: energy supply security, economic sustainability, and concerns over global climate change. As President-elect Obama alludes, the transportation sector is at the crux of this dilemma: high fuel price volatility directly affects the health of our economy and economic competitiveness, roughly two thirds of our transportation fuels are imported, and transportation is responsible for more carbon dioxide (CO_2) emissions than any other end-use sector of our economy (34% of our total CO_2 emissions by end-use). [1,2]

Coal to Liquids (CTL) is a commercial process which converts coal into diesel fuel, producing a concentrated stream of CO_2 as a byproduct. Coupling the process with carbon sequestration is relatively inexpensive (adding only 7 cents per gallon to the Required Selling Price (RSP) of the diesel product) and results in *a fuel with appreciably less (5-12%) life cycle Greenhouse Gas (GHG) emissions than the average U.S. petroleum-derived diesel.* This latter finding is in contrast to an earlier, high level analysis by the Environmental Protection Agency (EPA) which found CTL to have life cycle GHG emissions above that of petroleum. *This diesel fuel is compatible with our current fuel distribution infrastructure, can be used directly in existing diesel vehicles, and would be economically competitive with petroleum-derived diesel when the crude oil price (COP) is equal to or above $86 per barrel (bbl),* based on a twenty percent rate of return, January 2008 costs, and a GHG emissions value of zero.

This same basic process can be used to leverage domestic and widely available biomass (non-food) resources. For example, a mixture of eight percent (by weight) biomass and ninety-two percent coal – *Coal and Biomass to Liquids (CBTL) – can produce fuels which are economically competitive when crude prices are equal to or above $93/bbl and which have 20% lower life cycle GHG emissions than petroleum-derived diesel.*

Increasing the percentage of biomass in the feed further reduces the life cycle GHG emissions of the fuel, but also increases capital and operating costs due to the higher cost of biomass feedstock and reduced economies of scale. Diesel produced in a biomass only – i.e. Biomass to Liquids (BTL) – only becomes economically competitive when the GHG emission value exceeds $130/mt CO_2 Equivalents (CO_2E) and does not result in greater reductions in net GHG emissions than if the biomass were used in a CBTL plant.

Based on these findings, it is anticipated that CTL and CBTL with modest biomass percentages (less than thirty percent by weight) would, as a part of the United States' energy portfolio, provide a balanced solution to the nation's transportation fuel dilemma, providing affordable fuels from domestic feedstocks, and enabling significant reductions in GHG emissions.

Furthermore, a national commitment to promote the use of CTL and CBTL would have a tremendously positive impact on the economy, creating skilled jobs and reducing the amount of money sent overseas for oil imports, valued at $326 billion in 2007 and between $400 and $500 billion in 2008. The production of domestic diesel would also improve the economic competitiveness of domestic industries by easing supply constraints associated with diesel fuel, thereby reducing overhead costs associated with high fuel costs. Should oil prices resume their upward trend, the benefits of CBTL to the nation could be enormous.

NOMENCLATURE

AGR	Acid Gas Removal
AGT	Acid Gas Treatment
ASU	Air Separation Unit
ATR	Auto-Thermal Reformer
bbl	Barrel
BEC	Bare Erected Cost
BFW	Boiler Feed Water
BPD	Barrels Per Day
BTL	Biomass to Liquids
Btu	British thermal unit
CBTL	Coal and Biomass to Liquids
CCS	Carbon Capture and Storage
CFB	Circulating Fluidized Bed
CH4	Methane
CMM	Coal Mine Methane
CMT	Constant-Maturities Treasury
COE	Crude Oil Equivalent
COP	Crude Oil Price
COS	Carbonyl Sulfide
CO_2	Carbon Dioxide
CO_2E	CO_2 Equivalents
CTL	Coal to Liquids
CW	Cooling Water
DB	Daily Barrel
DOE	Department of Energy
DSCR	Debt Service Coverage Ratio
ECN	Energy research Centre of the Netherlands
eGRID	Emissions & Generation Resource Integrated Database
EIA	Energy Information Administration
EISA	Energy and Independence & Security Act
EOR	Enhanced Oil Recovery
EPA	Environmental Protection Agency
FEED	Front End Engineering Design
FR	Forest Residues
FT	Fischer-Tropsch
GDP	Gross Domestic Product
GHG	Greenhouse Gas
GHGEV	Greenhouse gas emission value
GREET	Greenhouse Gases, Regulated Emissions, and Energy Use in Transportation
GWP	Global Warming Potential
HHV	Higher Heating Value
H_2	Hydrogen
H_2S	Hydrogen Sulfide

IGCC	Integrated Gasification Combined Cycle
IPCC	Intergovernmental Panel on Climate Change
IRROE	Internal Rate of Return on Equity
ISO	International Standards Organization
lb	Pound
LIBOR	London Interbank Offered Rate
LCA	Life Cycle Assessment
LCFS	Low-Carbon Fuel Standard
LCI	Life Cycle Inventory
LHV	Lower Heating Value
MDEA	Methyldiethanol Amine
mmb/d	Million Barrels per day
MMBtu	Million Btu
MPG	Mixed Prairie Grass
MW	Megawatt
NETL	National Energy Technology Laboratory
N_2	Nitrogen
N2O	Nitrous Oxide
NOx	Nitrogen Oxide
NPV	Net Present Value
O_2	Oxygen
ppmv	Parts Per Million Volume
psia	Pounds Per Square Inch Absolute
R&D	Research & Development
RD&D	Research, Development & Demonstration
RSP	Required Selling Price
SG	Switchgrass
Syngas	Synthesis gas
TPD	Tons Per Day
ULSD	Ultra-Low-Sulfur Diesel
WGS	Water Gas Shift
WTT	Well-To-Tank
WTW	Well-To-Wheels
wt %	Weight Percent

1. INTRODUCTION

This study evaluates the use of the United States' abundant domestic resources to address the concurrent strategic objectives of energy supply security, economic sustainability, and the mitigation of global climate change. Addressing these objectives in the transportation sector is of particular immediate concern based on the high level of petroleum imports for this sector and recent high oil price volatility which negatively impacts both the health of the economy and economic competitiveness. Moreover, the vast distributed nature of point sources of greenhouse gas (GHG) emissions within transportation, a sector accounting for over a third of the country's total emissions, renders emission reduction inherently difficult.

The indirect liquefaction of coal is a near-term pathway that allows these objectives to be achieved. This Coal to Liquids (CTL) process uses three existing technologies – carbon capture, gasification and Fischer-Tropsch (FT) synthesis – to convert coal to diesel fuel, producing a concentrated stream of carbon dioxide (CO_2) as a byproduct. In other words, carbon capture is already part of the process. The results of a detailed modeling effort by the National Energy Technology Laboratory (NETL) show that when coupled with carbon sequestration, the overall process produces a product that has *significantly less (5-12%) life cycle GHG emissions than the average U.S. petroleum-derived diesel. These fuels are economically competitive with petroleum-derived diesel when the crude oil price (COP) is at or above $86 per barrel (bbl)* (based on a twenty percent rate of return, in January 2008 dollars, carbon price is zero). When carbon prices increase, the Required Selling Price (RSP) falls.

This same process can be used to leverage domestic biomass (non-food) resources. When an 8 percent by weight (8wt%) biomass feed is co-gasified with coal, the resulting process – Coal and Biomass to Liquids (CBTL) with carbon sequestration – can produce fuels which are economically competitive at crude prices above $93/bbl and which have 20% lower life cycle GHG emissions than petroleum-derived diesel.[1]

Based on these findings, CTL and CBTL with modest amounts of biomass (less than 30% by weight) would provide a balanced solution to the nation's energy dilemma, producing affordable fuels from domestic feedstocks and enabling significant reductions in GHG emissions. Furthermore, a national commitment to promote the use of CTL and CBTL would, at large scale, greatly benefit the economy, creating highly technical jobs and reducing the amount of money sent overseas for oil imports, estimated at $326 billion dollars in 2007 and between $400 and $500 billion in 2008.

1.1. Strategic Significance: An Energy Strategy Dilemma

The United States of America – like many other oil-importing countries in the world – is currently faced with competing strategic objectives related to energy, each with its' own set of significant challenges:

- Energy supply security: A lack of secure, reliable and adequate supplies of energy, combined with a relentless growth in imports from a world market that is heavily dependent on unreliable or potentially unstable sources of supply,
- Economic sustainability: A widespread concern for the health and sustainability of the nation's economy and standard of living, with the combination of high and volatile prices and import dependency sapping the nation's competitiveness,
- Climate Change: A growing consensus regarding the need to widely transform the nation's energy industries, infrastructure and consumption patterns to dramatically reduce GHG emissions, in an attempt to reduce the potential impacts of energy use on climate change.

A fundamental impediment for achieving these goals simultaneously is the difficulty that resolving individual challenges often serves to greatly exacerbate the others. For example, unconventional energy alternatives that could be used to supplement U.S. energy supplies, such as oil-sands and shale-oil have comparatively high GHG emissions associated with their production and use. Similarly, certain low GHG fuel sources have high production prices or limited availability, thereby yielding climate change benefits to the detriment of economic sustainability and energy supply security. This dilemma is described in Figure 1-1: moving too far in any direction has negative impacts on another goal.

The transportation sector and the fuels used therein are crucial components of these challenges: transportation fuels are our largest single area of oil consumption (14 million barrels per day (mmb/d) or 68% of total petroleum consumption) and the sector is responsible for 34% of all CO_2 emissions in the United States, making it the largest end-use sector emitter at 2,014 million metric tons of CO_2 in 2007.[2] [1,2,3]

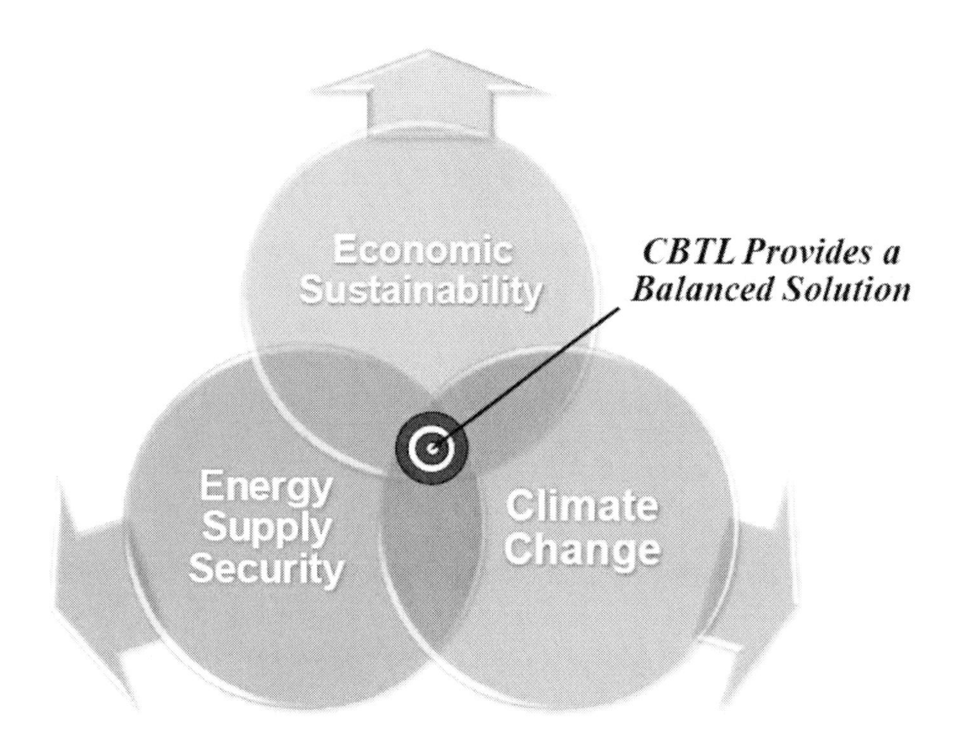

Figure 1-1. Solving the Energy Strategy Dilemma.

1.2. A Balanced and Attractive Solution

Although this threefold energy strategy dilemma may appear to be intractable, CTL/CBTL with carbon capture and sequestration (CCS) addresses these concomitant concerns, producing affordable, low-carbon diesel fuel from domestic resources, and therefore providing a balanced and elegant solution to the dilemma. More specifically, the key advantages of this technology in supporting each of these energy strategies are:

- Energy supply security:
  - o Addresses the transportation sector, which accounts for two-thirds of total oil demand [3]
  - o Uses coal and biomass: abundant and domestically available feedstocks
  - o Synergistic use of coal and biomass in CBTL:
    - Better economic and GHG benefits than biomass alone
    - Most cost-effective use of limited biomass resources
  - o Near-term pathway
    - Technology is commercial and ready for deployment now
    - Fuels produced work as a drop in replacement to diesel fuel
- Economic improvement and sustainability:
  - o Economic feasibility:
    - Feasible at crude oil prices above $86/bbl (twenty percent rate of return, carbon price is zero)
    - At carbon prices greater than zero, required crude oil feasibility price falls
  - o New highly technical industry which results in significant job creation
  - o Large-scale potential to address oil import dependency
  - o High profitability under recent forecasts of future oil prices
- Climate Change:
  - o Relative to a petroleum-derived diesel produced in a standard U.S. refinery:
    - GHG emission reduction of *5% to 12%, without biomass*
    - GHG emission reduction up to *75%, with combinations of up to 30% biomass*[3,4]
    - o Effectively addresses supply-side transportation sector emissions
    - A near-term and significant technical achievement
  - o Economically advantageous early opportunity for CCS demonstration projects
    - Incremental cost of CCS is low due to inherent CO_2 separation

1.3. Key Results

The key findings of our report may be grouped into three overlapping areas: the attractive low carbon profile at reasonable cost, the synergistic benefit of combining coal and biomass into the same process, and the potential economic benefits of a large scale CTL/CBTL industry.

1.3.1. Affordable, Low-Carbon Diesel Fuel

CTL/CBTL/Biomass to Liquids (BTL) plants can produce diesel fuel that has life cycle GHG emissions that are significantly reduced from the life cycle GHG emissions of petroleum-derived diesel fuel. Greater reductions are achieved as the percentage of biomass in the feedstock is increased and/or more aggressive carbon capture technologies are utilized (e.g. the CBTL with aggressive CCS cases). However, both of these configuration changes make the fuel more expensive to produce, thereby increasing the price at which it must be sold at (i.e. the Required Selling Price (RSP)) to achieve economic feasibility.[5]

Table 1-1 illustrates the proportional relationship between life cycle GHG emissions reductions and RSP.

The three columns, moving from left to right, represent the no carbon sequestration case (the CO_2 is captured but vented), simple CCS (CCS), and the more aggressive CCS configuration (CCS+ATR). The rows represent varying amounts of biomass with the top row consisting of no biomass (CTL) and the bottom row consisting of 100% biomass (BTL). The background colors of the cells represent the crude oil price required for economic feasibility, assuming a 20 percent rate of return and a GHG emissions value (i.e. carbon price) between \$0 and \$45/mtCO_2E. The cells with a green background are feasible at crude oil prices between \$80/bbl and \$100/bbl, while the yellow background corresponds to feasibility between \$100/bbl and \$120/bbl, and the red background represents feasibility when crude oil prices are at or between \$170/bbl and \$240/bbl.

The CTL with CCS configuration and the 8wt% to 15wt% CBTL with CCS configurations offer what might be the most pragmatic solution to the dilemma described in Figure 1-1: GHG emission reductions which are significant (5 to 33% below the petroleum baseline) at diesel RSPs that are only half as much as using biomass alone (\$2.56 to \$2.82/gal compared to \$6.45 to \$6.96/gal for BTL).[6] These CTL/CBTL with CCS options are economically feasible when crude oil prices are \$86 to \$95/bbl at a GHG emission value of \$0/mtCO_2E.

Table 1-1. Life Cycle GHG Emissions of CBTL Plant Compared to 2005 Petroleum Baseline

wt% Biomass	Carbon Capture Configuration			
	No CCS	CCS	CCS+ATR	
	Life-Cycle GHG Emissions Relative to Petroleum-Derived Diesel			
0	+147%	-5%	-12%	crude oil from \$80 to \$100/bbl*
8		-20%	-28%[1]	
15		-33%	-42%	crude oil from \$100 to \$120/bbl*
30		-63%	-75%	
100	-109%	-322%	-358%	crude oil from \$170 to \$240/bbl*

Economic feasibility point, assuming 20% IRROE & GHG Emissions Values ranging from **\$0/mtCO_2E to **\$45/mtCO_2E***

[1] The 8wt% CBTL with CCS+ATR point is interpolated between 0wt% and 15wt%.

While these configurations provide what may be the most balanced solution from an energy strategy perspective, CTL/CBTL/BTL configuration options can also be evaluated on a basis that assumes GHG emissions must be accounted for economically, such as if a cap and trade policy were implemented. In this scenario — described in Chapter 4 — any life cycle GHG emissions associated with the fuel would result in an operating cost (thereby increasing the RSP of the fuel) or, in the case where the fuel usage has negative GHG emissions, a revenue stream (based on GHG credits or offsets which can be sold).

Tables 1-2 and 1-3 summarize which CTL/CBTL/BTL plant configurations would be economically preferred at different levels of GHG emission valuations, i.e. which option would produce the fuel with the lowest RSP. As shown in Table 1-2, for GHG emissions values up to \$5/mtCO_2E, CTL without CCS is the economically preferred option, while for emissions values between \$5/mtCO_2E and \$101/mtCO_2E, CTL with CCS is preferred. These options are economically viable when the crude oil price is between \$74/bbl and \$85/bbl (the exact price varies based on the GHG emission value).

Table 1-2. Economically Preferred CBTL Configurations at Various GHG Emission Valuations without a LCFS[7]

Without a Low Carbon Fuel Standard		
GHG Emission Value, Jan-08 \$/mtCO_2eq	Preferred CBTL Alternative (lowest cost producer)	Crude Oil Price Required for Parity Between CBTL Diesel and Petroleum-Derived Diesel
0 to 5	100% Coal, no CCS, 50k bpd	\$84 to \$85 per bbl
5 to 101	100% Coal, CCS, 50k bpd	\$85 to \$74 per bbl
101 to 138	15 wt% SG, CCS, 50k bpd	\$74 to \$66 per bbl
138 and higher	100 wt% SG, CCS+ATR, 5k bpd	\$66 and lower

Table 1-3 varies from Table 1-2 in that it assumes that a Low Carbon Fuel Standard (LCFS) is in place that prohibits the production of certain fuels based on life cycle GHG emissions. In this case, fuels are required to exhibit emissions that are 20% below the petroleum baseline. In this scenario, none of the CTL options are permitted (CTL with CCS+ATR is still only 15% below the petroleum baseline) and CBTL options with 8wt% biomass and 15wt% biomass are the economically preferred configurations for GHG emission values up to \$58/mtCO_2E and \$138/mtCO_2E, respectively. These options are viable when crude oil is equal to or greater than \$66/bbl or \$93/bbl, depending on the GHG emissions value.

1.3.2. Synergistic Use of Coal and Biomass

This analysis reveals that BTL is very costly. For example, given a fixed amount of biomass available in an area adjacent to the CBTL or BTL plant, the use of that biomass in conjunction with coal results in greater investment returns, more fuels produced, and greater overall GHG emission reductions compared to using biomass alone. These effects are the result of both the large economy of scale achievable with coal (compared to biomass alone), and the improved thermal conversion efficiency when coal and biomass are co-converted to fuels compared to the efficiency due to the use of a circulating fluidized-bed gasifier in BTL systems. These results are discussed in Chapter 5 and point to CBTL as the best way to leverage biomass for fuels production.

Table 1-3. Economically Preferred CBTL Configurations at Various GHG Emission Valuations with a LCFS7

Under a 20% Low Carbon Fuel Standard		
GHG Emission Value, Jan-08 $/mtCO$_2$eq	Preferred CBTL Alternative (lowest cost producer)	Crude Oil Price Required for Parity Between FT Diesel and Petroleum-Derived Diesel
0 to 58	8 wt% SG, CCS, 50k bpd	$93 to $83 per bbl
58 to 138	15 wt% SG, CCS, 50k bpd	$83 to $66 per bbl
138 and higher	100 wt% SG, CCS+ATR, 5k bpd	$66 and lower

1.3.3. Economic Benefits

CTL/CBTL plants and a potential CTL/CBTL industry have a number of advantageous economic benefits. These benefits, discussed in Chapter 6, include the likelihood of large and growing earned profits. When diesel prices, and the corresponding crude oil prices, rise above the level required for CTL/CBTL to be economically feasible, economic profits will escalate as well. Thus, if world oil prices are above $90/bbl, a scenario to which the world may quickly return as it recovers from the current global recession and financial crisis, every barrel of CTL/CBTL would produce substantial economic benefits. These benefits include the moderation of world oil prices, the retention of economic rent, possible amelioration of the trade deficit, and extensive domestic job creation, on the order of 150,000 jobs per million bbls of CTL/CBTL. Over the period 2010-2030, the net present value (NPV) of ramping up to a 3 million bpd industry could range from $200 billion to $700 billion, in 2008 dollars.

1.4. About this Study

This study evaluates the performance and cost of eleven different CTL/CBTL/BTL plant configurations in order to identify a balanced solution to the nation's energy strategy dilemma. The entirety of the analysis leverages the extensive experience of NETL in the gasification of carbonaceous feedstocks, large-scale energy conversion, indirect liquefaction via the FT synthesis process, and carbon capture and storage, including in-depth work with the Carbon Sequestration Regional Partnerships over the last decade. The scope of the study is limited to one type of biomass and one type of coal, with additional studies to follow which consider alternate feedstocks, plant configurations, and plant locations.

The plant configurations evaluated in this study focus on optimally producing liquid fuels while dramatically reducing CO_2 emissions, creating scenarios and sensitivities for addressing the conflicting priorities of transportation GHG mitigation and energy supply security. While it is noted that CTL/CBTL/BTL plants may be designed to produce significant amounts of excess electric power based on market conditions, all the configurations in this study were designed to produce little, if any, power in excess of what the plant itself needs to operate. This also allows the economic analysis to be uninfluenced and uncomplicated by difficult assumptions regarding:

- crediting and allocation of life-cycle GHG emissions between electric power and fuels, and
- valuation of low-carbon electric power under future GHG regulation, including its price ratio with liquid fuel prices.

Additionally, no credit is taken for soil root carbon, i.e. the accumulation of carbon in the soil and roots of energy crops, as there is some question as to the appropriate accounting method which should be used for this carbon. This report therefore may significantly understate the potential GHG benefits of biomass usage. This has very little effect on the overall economic findings, e.g. which option is preferred at what carbon price, but could result in CBTL fuels which produce net zero GHG emissions with as little as 35-40 wt% biomass.

The report is structured as follows:

- Chapter 2 describes the CTL/CBTL/BTL process, commercial readiness of the associated technologies, and the design strategy used and the plant configurations evaluated in this study;
- Chapter 3 discusses the life cycle GHG footprint of CTL/CBTL/BTL produced diesel fuel, comparison of these fuels to petroleum-derived diesel, and provides a detailed look at GHG emissions from the CTL/CBTL/BTL process;
- Chapter 4 details the overall capital cost estimates and illustrates the economic feasibility of CTL/CBTL/BTL;
- Chapter 5 establishes the preference for combined use of coal and biomass; and
- Chapter 6 outlines the potentially large economic benefits of a CTL/CBTL/BTL industry.

2. Converting Coal and Biomass to Diesel Fuel: The FT Process

The nation is looking towards transportation fuels produced from renewable resources, such as biomass, as a means of addressing the joint challenges of energy security and climate change. This goal is being pursued through numerous legislative avenues as well as extensive research and development into both new technologies and efforts at improving burgeoning technologies which are not currently competitive.

This chapter describes an existing and proven technology for producing diesel fuel from both biomass and/or coal: Coal and Biomass to Liquids (CBTL), Coal to Liquids (CTL) and Biomass to Liquids (BTL).

CTL/CBTL/BTL using Fischer-Tropsch (FT) synthesis is a near-term solution for diesel fuel production: the technology has been in commercial use since the 1930s and the fuel produced can be used in today's fueling infrastructure.[8] It can also be used to produce *fuels which have a life cycle GHG emissions profile which is less than that of petroleum-derived diesel* by coupling the process with carbon sequestration. This can be done *at a very small incremental cost (less than $0.10/gallon of diesel fuel)* due to the nature of the CBTL process, which produces a pure stream of CO_2 as part of the process.[9,10]

The co-gasification of coal and biomass (CBTL) uses coal to overcome key challenges that face the use of biomass as a feedstock. These include supplying additional feedstock to enable larger scale plants to be built (economies of scale improve plant economics) and preventing plant downtimes if biomass is not available.

At the same time, coal benefits from biomass, which is a renewable resource (thereby providing a sustainable energy source) that "recycles" carbon from the atmosphere, a substantial benefit in terms of climate change.

All of these aspects of CBTL make it an attractive solution for producing affordable, low-carbon diesel fuel from domestic resources, thereby enhancing energy supply security, promoting economic sustainability and addressing climate change issues associated with the transportation sector.

The following sections describe the CTL/CBTL/BTL process, the technological readiness of different technologies used in the plant, and the specific plant configuration and feedstock pairs examined.

CBTL: Diesel Fuel from Coal and Biomass

CBTL is a generic term describing the conversion of coal and biomass to liquid fuels. This study evaluates a specific CBTL process – indirect liquefaction with FT synthesis – as a means to convert carbonaceous materials (e.g. coal and biomass) into diesel fuel.

The FT process falls into the category of "indirect liquefaction" because the feedstock (coal and/or biomass) is first broken down into building block molecules – carbon monoxide (CO) and hydrogen (H_2) – via gasification, then this "synthesis gas"(syngas) is converted into liquid hydrocarbons via FT catalytic synthesis, a large percentage of which can be used to produce premium diesel or jet fuels. This differs from other direct liquefaction and pyrolysis technologies that liquefy the coal and biomass directly by cracking large molecules and adding H_2, rather than first producing a clean gas that is then converted to liquids.

FT diesel fuel can be blended with or used as a drop-in replacement for petroleum-derived diesel that is compatible with today's infrastructure, cars, and other end-uses. Additionally, it is superior in quality to petroleum-derived diesel as it is essentially free of sulfur, lower in life cycle GHG emissions (if carbon containment techniques are used), and produces less particulate matter during combustion. [5]

2.1. CBTL Process Description

Many different options exist for the design and configuration of a CTL/CBTL/BTL plant and these options can result in wide variations in the plant cost and performance. The plants described here are designed for maximum diesel fuel production (production of co-products such as electricity is minimized) and are evaluated at various levels of both CO_2 capture and biomass percentage in the feed.

Regardless of size, overall configuration, and feedstock, the conceptual plants analyzed in this study all have certain process units in common. This section provides a detailed look at the unit processes in each of the three basic CTL/CBTL/BTL plant configurations evaluated: no GHG mitigation ("without CCS"), "simple CCS", and "aggressive CCS".

The "without CCS" and "simple CCS" plant configurations are functionally equivalent as CO_2 is captured in both cases. The key difference is that in the "simple CCS" case, captured CO_2 is compressed, transported and stored in a geologic formation whereas in the other case it is merely vented to the atmosphere. Figure 2-1 is a simplified block flow diagram which describes this plant configuration. The extra equipment for compression, transport, and storage, combined with the fund for CO_2 monitoring, only constitute 4% of the total capital cost of the plant, resulting in a small incremental cost to add carbon sequestration to the plant, as will be described in greater detail in Chapter 4.

The "aggressive CCS" case is a CTL/CBTL/BTL plant configured for more aggressive levels of CO_2 capture (>95%). This is achieved through the use of an Auto-Thermal Reformer (ATR), an additional Water Gas Shift (WGS) unit, and a revised recycle stream. Figure 2-2 is a simplified block flow diagram which describes this plant configuration, with the process changes from the "simple CCS" configuration denoted in red.

The process design choices and specific cases evaluated are described in greater detail below in the Sections 2.4 and 2.5. This includes details on the choice of switchgrass (SG) as a representative biomass feedstock, biomass availability and how that relates to plant production capacity, GHG emissions reduction strategies, and the focus on fuels production.

2.1.1. Feedstock Processing and Drying

Both feedstocks – coal and biomass – must be prepared for conversion by grinding and drying. The switchgrass feedstock also undergoes some preparation prior to arrival at the CBTL plant: it is cut, field dried to 15wt% moisture, then baled at the collection site where it is stored until needed at the plant. Bales of switchgrass are transported by truck, and at the plant, a de-baler breaks up the bales into loose grass and uses waste heat from this equipment to dry the biomass to a nominal 10% moisture (by weight) as it is fed into the grinding and final drying process. Biomass is more reactive than coal, and therefore does not have to be ground as fine: grinding to a size of one millimeter or less is required in order to ensure proper feeding. It is dried to 5% moisture (by weight) using driers fired with FT tail gas, prior to feed into the gasifier. See Section 2.4 for more information on the choice of switchgrass as a biomass feedstock and information on cultivation, harvest, and processing prior to the plant gate.

Coal is transported to the plant via rail and is crushed and ground to a size distribution which is 17 percent less than 200 mesh. Coal is also dried to 5% moisture (by weight) prior to feed into the gasifier.

CBTL Terminology

Coal to Liquids (CTL) – A plant which converts coal to liquid transportation fuels, in this case via FT synthesis.

Coal and Biomass to Liquids (CBTL) – Similar to CTL, this plant converts both coal and biomass to fuel.

Biomass to Liquids (BTL) Similar to CTL/CBTL, this plant only uses biomass as a feedstock for fuel production.

Carbon Capture and Storage (CCS) – The capture, transport, and long-term storage of CO_2 to reduce GHG emissions and the climate change impact of a process. The CCS cases evaluated here

Simple CCS ("CCS cases") – CBTL plant with a simple CCS system in which greater than 91% of the CO_2 produced by the plant is captured.

Aggressive CCS ("CCS+ATR cases") – CBTL configured for more aggressive levels of CO_2 capture (>95%).

This is achieved through the use of an Auto-Thermal Reformer (ATR) (discussed below).

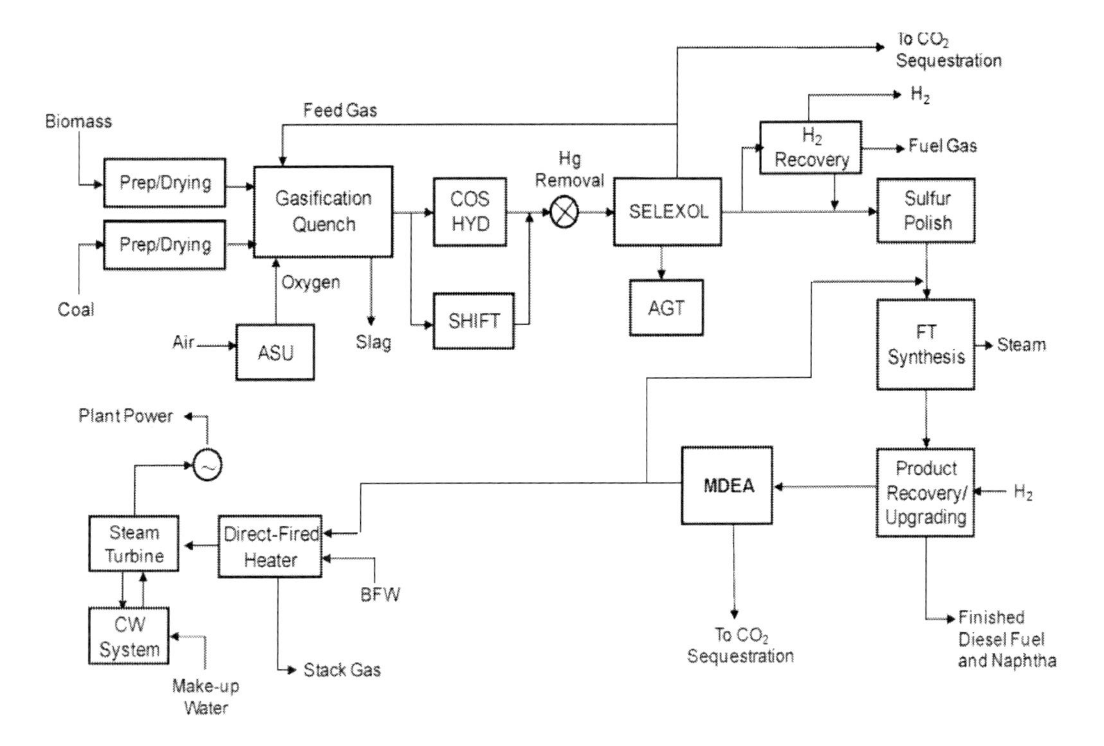

Figure 2-1. Simplified Block Flow Diagram for the CBTL Plant Equipped for Simple CCS.

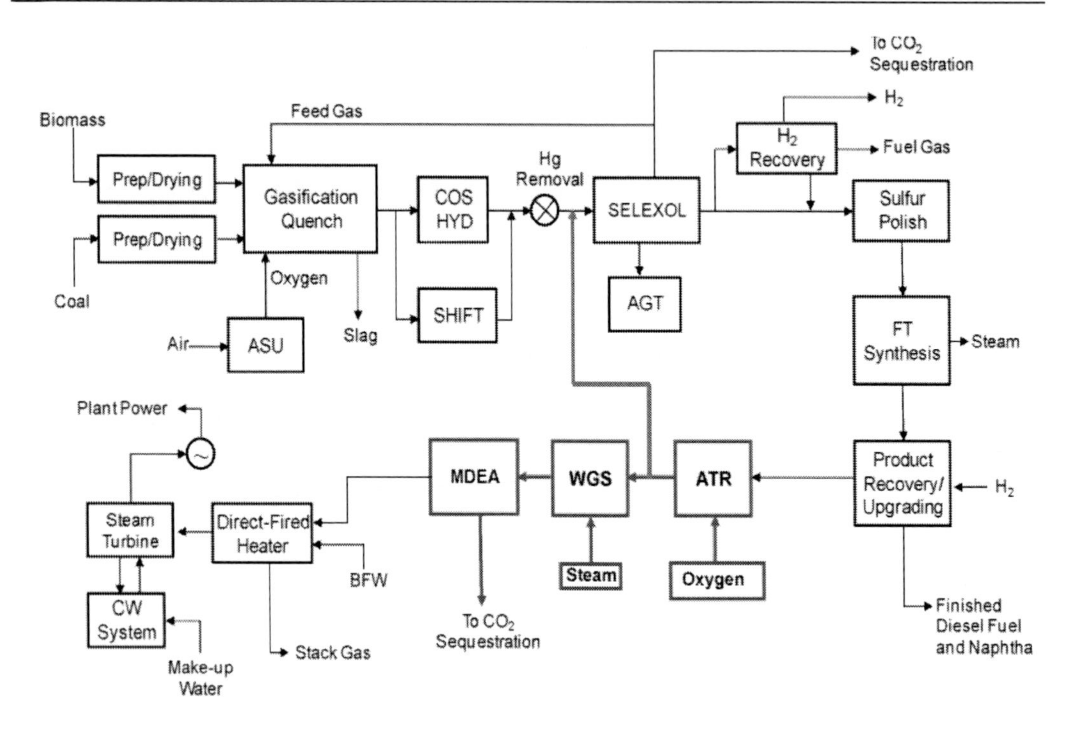

Figure 2-2. Simplified Block Flow Diagram for the CBTL Process Equipped for Aggressive CCS11.

2.1.2. Gasification

The indirect liquefaction process requires solid feedstocks to be gasified to synthesis gas (syngas) before they can be converted to a liquid fuel. As part of the overall study, a review of gasification technology suitable for CBTL processes was performed and the gasifiers and operating conditions used in the conceptual plant designs were based on recommendations from this work. [6] Oxygen-blown gasifiers were used in all process configurations. This results in significantly reduced equipment sizes throughout the plant and therefore dramatically reduces the overall plant costs.

2.1.2.1. Gasification of Coal and Coal/Biomass Mixtures

In CTL and CBTL cases, a single stage, dry feed, entrained-flow gasifier was used to gasify the coal and/or biomass. This type of gasifier was chosen due to operating experience in co-firing biomass and the advantage that it produces no tars and a minimal amount of methane (CH4) (which does not react in the FT synthesis process).

The gasifier is of the slagging type and a direct contact water quench spray system is used to cool the syngas exiting the gasifier. The quench also removes particulate matter and contaminants not removed in the slag. However, because the ash from biomass is rich in calcium oxide, it is difficult to melt even at the high gasifier operating temperature (2600°F) and additional fluxing agents may be required to obtain acceptable slag properties. It is assumed in this study that the gasifier design has to be modified to include the two separate feed systems and dedicated biomass burners.[12]

2.1.2.2. Biomass Only

The BTL cases use a circulating-fluidized bed (CFB) gasifier for the biomass. This system operates at lower temperatures and will therefore produce more light gases, like CH_4, and tars from the breakdown of the biomass, requiring additional processing to take place downstream of the gasifier. The CFB gasifier was chosen as it is generally accepted to be an appropriate gasifier for pure biomass streams, allowing: larger feedstock particle sizes - important due to the high energy requirements associated with grinding fibrous biomass - lower temperature operation (acceptable based on biomass reactivity), and at scales more suitable to generalized biomass feedstock availability. However, the operating experience with these gasifiers tends towards air-blown operation as opposed to the oxygen-blown systems evaluated in this study. The CFB gasifier is also slightly less efficient than the entrained-flow gasifiers used in the CTL/CBTL cases due to the increased steam load of the process.

2.1.3. Air Separation Unit

The oxygen for gasification is provided by a conventional cryogenic Air Separation Unit (ASU). This unit takes uses a cryogenic distillation column to separate air into oxygen (O_2) (95% purity) and nitrogen (N_2).

2.1.4. Gas Cooling, Raw Water Gas Shift, COS Hydrolysis, and Mercury Removal

The syngas stream leaving the gasifier quench section is split into two streams. The first stream is sent to a raw WGS reactor where water is reacted with the syngas to generate additional H_2 via Equation 1. This H_2 is required for both the FT reactor and the product upgrading section of the plant.[13] The other portion of the syngas is sent to a carbonyl sulfide (COS) hydrolysis unit where the COS is hydrolyzed to hydrogen sulfide (H_2S), a form which can be removed downstream in the acid gas removal (AGR) unit. (COS is also hydrolyzed in the WGS reactor.) While splitting the stream in this fashion is not strictly necessary – it is possible to operate the WGS reactor in such a way that the desired level of H_2 is produced – it does result in reduced costs and is the generally accepted process configuration for CBTL plants.

Water Gas Shift: $$CO + H_2O \leftrightarrow CO_2 + H_2 \qquad (2\text{-}1)$$

The two streams are then combined and cooled in gas coolers before being sent to activated carbon filtration for removal of mercury. The cooled gas is then sent to a two-stage SelexolTM unit for removal of H_2S and CO_2.

2.1.5. Acid Gas Removal

The SelexolTM unit is used for the selective removal of H_2S and for bulk removal of CO_2. The H_2S produced by this selective absorption is suitable for feeding to a Claus-type unit for acid gas treatment (AGT) and recovery of elemental sulfur. In the CCS cases, the CO_2 is sent to drying and compression; otherwise, it is vented to the atmosphere.

2.1.6. Sulfur Polishing

Depending on operating conditions, the syngas exiting the SelexolTM unit still contains about 1-2 parts per million of H_2S on a volume basis (1-2 ppmv). This quantity of H_2S is still too great to feed to the sulfur sensitive iron-based catalysts in the FT synthesis process. To remove this residual H_2S, zinc oxide polishing reactors are used. The zinc oxide reacts with the H_2S to form solid zinc sulfide. The product gas leaving the polishing reactor contains less than 0.03 ppmv H_2S.

2.1.7. Hydrogen Recovery

A portion of the clean syngas leaving the AGR unit is sent to the H_2 recovery unit where sufficient H_2 is separated and purified for use in the FT upgrading section of the plant. This H_2 is required for hydrotreating and hydrocracking. The H_2 separation system chosen for this study is the combination of a membrane unit and a Pressure Swing Adsorption (PSA) unit.

2.1.8. Fischer-Tropsch Synthesis

The clean syngas from the sulfur polishing reactor is sent to the FT section of the plant. The syngas is heated to about 400ºF and fed to the bottom of the FT reactors, which operate in the 360-480ºF temperature range. The gas bubbles up through the reactors that are filled with liquid hydrocarbons in which are suspended fine iron-based catalyst particles. Reaction heat is removed via heat exchange tubes within the reactors. The liquid medium enables rapid heat transfer to the heat exchangers which allows high syngas conversions in a single pass through the reactor, and conversions of up to 80 percent per pass can be obtained. In addition to the FT catalysis reaction, a WGS reaction also occurs in the presence of the iron catalyst, resulting in the conversion of CO and water to H_2 and CO_2 throughout the reactor, increasing the H_2 to CO ratio of the syngas.

Fischer-Tropsch Synthesis

The FT process creates what can be considered "synthetic" liquid fuels such as diesel and jet fuels from carbonaceous feedstocks. The solid feedstock, such as coal or biomass, is first broken up into CO and H_2 by gasification and gas cleaning and then reacted with a catalyst to form hydrocarbons of various lengths, the majority of which can be converted into transportation fuels. This differs from conventional petroleum refining where carbon chains are broken down into shorter chains to form fuel. Basic molecular building blocks are formed that result in fuels free of the sulfur and aromatic compounds found in petroleum-based fuels.

To do this, syngas is put into contact with a catalyst such as iron or cobalt, which allows the chemical conversion shown in Equation 2 to take place. By using different catalysts, different H_2 to CO ratios, or operating the reactor at different temperatures, different hydrocarbon products will be formed.

$$(2n+1)H_2 + nCO \rightarrow CnH(2n+2) + nH_2O \qquad (2\text{-}2)$$

The FT reactor used in this study is a low temperature (360-480ºF), slurry phase reactor which contains an iron catalyst. This reactor design and operating configuration are optimized for the production of long carbon chain hydrocarbons that can be selectively hydrocracked into diesel fuel and jet fuel, along with the minimization of oxygenates.[14] Slurry reactors also give a higher conversion per pass because of their superior heat transfer characteristics. Iron is used as catalyst because it is less expensive than cobalt and readily obtained in the U.S.

The immediate products from the FT reactor system are: (1) a tail-gas containing CO_2, unreacted H_2 and CO, and light hydrocarbon gases (C4 and below), (2) a liquid stream containing medium length hydrocarbons, and (3) a wax stream containing long chain hydrocarbons.

The tail-gas undergoes additional processing downstream before being split into a recycle stream and a stream to be combusted, as will be described below. A maximum of 80% of the tail gas can be used in the recycle stream based as higher recycle rates will result in unnecessarily large FT reactor units due to the buildup of extra syngas in the FT reactor loop.

The liquid and wax streams are sent to the product recovery/upgrading sub-system. In this area, the H_2 produced in the H_2 recovery system upstream of the FT reactors is used in hydrocracking the wax and hydrotreating the raw to produce a diesel fraction and a naphtha fraction.

2.1.9. FT Tail-Gas Processing (Aggressive CCS Cases)

Half of the cases enabled for CCS also utilize ATR and WGS reactors to attain increased levels of CO_2 capture, thereby further reducing the GHG footprint of the FT diesel fuel.[15] In these cases, the light hydrocarbons in the tail gas are partially oxidized to CO,, producing H_2 as a by-product and making it possible to capture carbon which would otherwise be combusted and emitted as CO_2. This partial oxidation occurs in the ATR, and generates steam for use in process units such as the WGS reactor.

The reformed tail gas is then split into two streams: (1) a stream of H_2, CO, and CO_2 that is recycled through the upstream Selexol system and FT unit in order to increase liquids yield, and (2) a stream that is shifted in a WGS reactor in order to convert any CO species to CO_2, thereby enabling CO_2 capture.

After the latter stream is shifted, it is processed in a methyldiethanol amine (MDEA) unit to remove the CO_2, resulting in a H_2 rich gas. This tail-gas is then sent to the power generation block for combustion.

A standard MDEA unit with a single CO_2 absorber and solvent regenerator is used for this unit operation. Captured CO_2 is compressed for transport and storage, as described below.

2.1.10. FT Tail-Gas Processing (Simple CCS Cases)

In the "simple CCS" cases the FT tail gas is processed in a MDEA unit to remove the CO_2 from the tail gas. CO_2 is removed in order to reduce the volume of the recycle stream (and therefore the size and cost of the FT reactor) and to improve the heating value of the tail gas to be combusted. After the FT tail gas is processed in an MDEA unit, it is split into two streams, one of which is sent to the power generation block for combustion and the second of which is a recycle stream. This recycle stream is then recycled to just upstream of the FT reactor.

A standard MDEA unit with a single CO_2 absorber and solvent regenerator is used for this unit operation. Captured CO_2 is compressed for transport and storage, as described below.

2.1.11. FT Tail-Gas Processing (non-CCS Cases)

The FT tail-gas in the non-CCS is processed in a fashion identical to that described in the "simple CCS" section (Section 2.1.10) except that captured CO_2 is vented to the atmosphere (as opposed to compressed for transport and storage).

2.1.12. CO_2 Compression, Transport, and Storage (CCS Cases only)

In cases enabled for CCS, CO_2 captured in the plant is dried and compressed for pipeline transport to 2,200 pounds per square inch absolute (psia), at which point it is a supercritical fluid. A pipeline length of 50 miles is assumed and the pipeline diameter is specified such that the CO_2 pressure is 1,200 psia at the pipeline destination, providing a ten percent safety margin above the critical-point. This design removes the need for recompression stages.

Transported CO_2 is injected into a saline formation for long-term storage with provisions for 80 years of monitoring to ensure the CO_2 remains in place.

The costs associated with each CCS stage – compression through monitoring – are included in both the selling price of the fuel and the capital and operating costs reported throughout this document. These costs represent approximately 4% of the overall capital costs, and therefore do not have a dramatic effect on the RSP of the final diesel fuel product.

2.1.13. Power Generation Block

The tail gas is combusted in a direct-fired superheater to create steam for power generation. This steam is used to drive the steam turbine to generate the electric power for the plant. Enough gas is combusted to provide power for all of the equipment in the plant and in some cases a small amount of excess power (less than ten percent of the total plant parasitic power draw) is sold to the local electricity grid.

2.1.14. Balance of Plant (BOP) Units

The conceptual design included materials and equipment for on-site tank product storage, plant water systems (cooling towers, boiler feed water (BFW), waste water treating, storm water handling, and fire water requirements), electrical transformers and plant power distribution facilities, and instrumentation and control systems.

2.2. Product Mix

The product mix produced by the FT reactor is 70% by volume FT diesel and 30% by volume naphtha. The FT diesel fuel contains no sulfur and is completely fungible with petroleum-derived fuels: it is ready for use in existing vehicles without any modification; can be blended with petroleum-derived diesel to improve the cetane content of that fuel; and can be transported and stored using the existing infrastructure. Furthermore, the literature reports that FT diesel has several advantages over petroleum-based diesel, including burning cleaner with fewer particulate emissions [5]. The zero sulfur fuel also allows lean Nitrogen Oxides (NOx) catalysts to be used in vehicle exhaust emission cleanup systems.

The naphtha generated by the plant is sold as a chemical or gasoline feedstock. The end use is likely to vary from plant to plant, based on current market conditions and proximity to chemical plants or refiners. Future analyses will investigate the refining and upgrading of naphtha to gasoline at the indirect liquefaction plant itself.

2.3. Technology Background and Readiness

This section reviews the maturity and history of each technology and identifies potential hurdles and opportunities for process optimizations.

2.3.1. Coal Gasification

Coal gasification is a mature technology which has been deployed throughout the world, including in conjunction with FT synthesis plants.

2.3.2. Biomass Gasification

The majority of existing biomass gasification systems are small scale, air-blown, and low temperature systems. Such gasifiers were deemed inappropriate for the BTL cases in this study for reasons including N_2 dilution in air-blown systems, gas cleanup issues, and problems with tar formation. Although considerable research has been performed on biomass gasification, a large, commercial scale biomass gasification plant has yet to be built which meets the operating requirements of this technology: high operating pressure, oxygen blown, and/or high temperature operation [7].

Unfortunately, there is very little data in the literature for high pressure biomass gasifiers. Because of the fibrous nature of most biomass sources, the material is very difficult to feed into a high pressure gasifier. Typical problems include clumping and bridging. Other technical hurdles also exist, such as the high slagging temperature of mineral content within biomass and the production of tars in low-temperature gasifiers often considered for use with biomass. These issues will require significant Research & Development (R&D) prior to deployment. Furthermore, research will be required associated with the gasification of biomass in O_2-blown systems, as required by the FT reactor and in order to drive costs down.

2.3.3. Co-Gasification of Coal and Biomass

Co-gasification of coal and biomass has been successfully demonstrated at both the Polk Power Plant in Tampa, Florida, and the NUON power plant in the Netherlands. The operating experience at Polk involved up to one and a half percent woody biomass (by weight) being co-gasified with bituminous coal. The biomass was preground to particles nominally 1/2" in size and did not present any major issues [8].

At NUON they successfully fed a mixture of 30 percent by weight of demolition wood and 70 percent coal to the Shell-dry-feed, high pressure, entrained gasifier. The wood was reduced to sawdust and had a particle size of 1 mm or less. Based on this experience it is assumed that it is possible to feed small wood particles of 1 mm size to a pressurized entrained gasifier using the existing lock hopper feed system. This assumption is supported by Energy research Centre of the Netherlands (ECN) in their report on entrained flow gasification of biomass [9].

These initial successes show the promise of co-gasification, but because they are limited to a few sources of biomass, limited duration tests, and various concentrations of biomass, further R&D is required before large-scale deployment. The major design issues which must be tackled include the development of high pressure biomass feed systems, gasifier optimizations for co-gasification at varied biomass feed rates, overcoming hurdles associated with feeding heterogeneous biomass types, design of biomass preprocessing systems and choice between biomass preprocessing at the plant or at the harvest site.

Therefore, while co-gasification can be considered a technology ready for long-term demonstration at large-scale, full implementation which allows the utilization of a wide range of heterogeneous biomass feedstocks will require additional Research, Development and Demonstration (RD&D).

2.3.4. Fischer-Tropsch Synthesis

FT synthesis is a commercial process which was utilized extensively in Germany through the end of World War II. It is currently being utilized commercially by SASOL and Petro-SA in South Africa, by Shell in Malaysia, and by SASOL in Qatar. The South Africa plants were deployed 25-30 years ago, and while SASOL has continued an active R&D program since then, no large scale facilities were built in the remainder of the 20th century. The 66,000 bpd Gas to Liquids plant currently under construction in Qatar represents the first large scale deployment of an FT synthesis plant by SASOL in 25 years.

The commercial nature of the process non-withstanding, R&D opportunities exist, including the development of better catalysts and improving the understanding of the FT process so that the product slate can be easily changed to match market trends. Improvement in these areas represents an opportunity to increase profitability of FT operations, as well affect the reduction of up front capital costs.

2.3.5. Systems Integration

In the twenty plus years since the SASOL units went into production, significant technological advances have been made, resulting in a number of opportunities for process efficiency and environmental improvements. Therefore, while CTL plants are considered commercial, best practices using new technologies have yet to be ascertained and significant improvements can be realized during systems integration.

2.3.6. Carbon Sequestration in Geologic Sinks

Currently, two major demonstration projects are sequestering an estimated two million tons of CO_2 in geologic formations: one for enhanced oil recovery (EOR) operations and one for the explicit sequestration of CO_2 to avoid CO_2 emissions taxes. The latter is the objective of the Sleipner CO_2 injection project, with one million tons of CO_2 being injected per year into a saline formation under the North Sea. This CO_2 is the byproduct of a natural gas production facility and injection has been occurring since 1996 and, coupled with extensive monitoring operations, has served as a large scale demonstration of how carbon sequestration can work. [10] EnCana's enhanced oil recovery project in Weyburn is another representative case where over a million tons are injected per year, resulting in over 20 million barrels of incremental oil produced and detailed modeling data of underground CO_2 flows in oil wells.[16] The CO_2 used in this project is produced from the conversion of coal into synthetic natural gas and other specialty chemicals at the Dakota Gasification Plant in Beulah, ND, and represents a good example of how CO_2 generated from coal fed processes can be used in an economically beneficial way [11]. These projects are in addition to the oil industry which uses CO_2 from natural deposits for EOR applications throughout the Southwest and Texas.

To date, geologic sequestration has not been widely deployed, predominantly due to the lack of legal requirements for reduction and lack of an economic driver to cover the large capital and operating costs of concentrating CO_2 from point sources such as power plants. The above examples – Sleipner and Weyburn – are unique in that relatively pure CO_2 was already being produced at the plant, and the costs, while significant, were offset by tax breaks. Both CTL and CBTL offer a similar opportunity, as the CO_2 generated by the plant is already concentrated into a pure stream and ready for compression and transport. This makes CTL and CBTL superb choices for carbon sequestration demonstration projects, as the marginal cost of CCS is very low.

2.4. Design Strategy

Many different options exist for the design and configuration of a CTL/CBTL/BTL plant and these options can result in wide variations in the plant cost and performance. This section describes the overarching design strategy used in this study.

2.4.1. Liquid Fuels Production Focus

The CBTL plants evaluated in this study are configured for the production of fuels (as opposed to other co-products, such as electricity). The decision to focus on liquid fuels production – and specifically diesel fuel – stems from a desire to simplify the analysis by eliminating the need to allocation production costs and GHG emissions to another significant byproduct.[17] Maximum fuels production is achieved by the use of a so-called "recycle configuration", in which syngas that is not converted to fuels in the FT reactor is recycled back into the process, maximizing the amount of carbon which is converted into diesel fuel.

2.4.2. GHG Emissions Reduction

A decision was also made to focus on reducing GHG emissions from the CTL/CBTL/BTL plant, based on current concerns related to climate change. As described in Section 2.1 above, two different CCS configurations were examined: a default, or "simple CCS" configuration, and a second "aggressive CCS" configuration in which equipment is added (at additional cost and performance penalty) in order to further reduce CO_2 emissions from the CBTL process.

The "simple CCS" is a low incremental cost option for CCS, as it is functionally identical to the "without CCS" cases: CO_2 is already captured within the CTL/CBTL/BTL plant as part of the process. The only difference is the addition of CO_2 compression, transport and storage capital and operating costs. *This option results in the capture of 91% or more of the CO_2 produced by the plant.*

The "aggressive CCS" plant configuration was developed in order to increase the level of CO_2 capture achievable in a CTL/CBTL/BTL plant, so that a lower GHG emissions profile fuel could be produced. The key to further reducing emissions is to aggressively remove as much carbon as possible from the fuel gas to be combusted for plant power generation. As a portion of the carbon in the fuel gas is in the form of light hydrocarbons (C2-C4), reductions can be achieved by converting this carbon to CO_2 so that it can be captured. This type of conversion is common in some CTL/CBTL/BTL process configurations, where light hydrocarbons are converted to CO (through partial oxidation) which is recycled to a point upstream of the FT reactors [13]. A similar technique was used in the "aggressive CCS" configuration, except that the light hydrocarbons are first partially oxidized to CO, then converted to CO_2 via WGS so that they can be captured in the MDEA unit.

These changes in the process result in the capture of more than 95% of the CO_2 produced by the plant although this additional level of capture incurs both an efficiency and cost penalty which in many cases makes this plant configuration not preferred economically.

2.4.3. Carbonaceous Feedstocks

The indirect liquefaction CTL/CBTL/BTL pathway offers a great deal of flexibility with regards to feedstock choice due to the extensive gas cleaning required to protect the FT catalyst. For example, coals which might be undesirable for power generation due to high sulfur content, low-heating values, or other undesirable characteristics can be used as a feedstock. Similarly, a wide variety of biomass types, ranging from herbaceous and woody biomass to agricultural waste (corn stover, bagasse, etc.) and construction wastes can also be used.

This study uses one feedstock of each type – bituminous coal and switchgrass – to evaluate the CTL/CBTL/BTL processes. These were chosen as representative feedstocks for a Midwest plant location. Other coal and biomass feedstocks will be evaluated in a later study.

2.4.3.1. Coal

Illinois #6, a high sulfur, bituminous coal is used as the coal feedstock in this study. The coal is mined underground using a combination of both conventional and long-wall mining techniques. If the mine is particularly gassy (high in coal mine methane), best coal mine methane (CMM) management practices are used to meet both Mine Safety and Health Administration (MSHA) mine safety requirements and to reduce CH4 emissions.[18] CMM is assumed to be combusted on site for use in mining operations due to variability such as CMM product quality and proximity to markets.

Coal is delivered to the CTL/CBTL plant by rail and contains 11.11% moisture (by weight).

2.4.3.2. Biomass

Only one biomass type, switchgrass, is evaluated in this study. Switchgrass is herbaceous biomass which can be grown throughout the United States including on degraded or marginal lands. This study assumes switchgrass is cultivated on these land types, which can affect feedstock cost and availability as described in Section 2.4.4.

Once harvested, switchgrass is left to dry in the field, resulting in a final moisture content of 15% (by weight). Of the cultivated crop, 15% (by weight) is assumed to have been lost during harvest. Field dried switchgrass is then collected then baled, covered with tarps, and stored on the ground in the field. A further 10% (by weight) of the switchgrass is assumed to be lost during storage due to biomass degradation. The bales are collected and transported by truck to the CBTL/BTL facility where they are processed in a de-baler, dried with waste heat to a moisture content of 10% (by weight),and sized for gasification.

2.4.4. Switchgrass Availability

A key issue surrounding the use of biomass as an energy feedstock is land use change, i.e. energy crops competing for lands used for food crops or causing non-croplands to be developed for cultivation, resulting in the release of stored carbon from these lands. One example of this would be the clearing of forests for additional cropland to be created. For the purposes of this study, it is assumed that no land use changes occurred as a result of biomass cultivation, and that the switchgrass used was cultivated only on marginal lands which are not suitable for food-crops (abandoned mine lands, etc.) or depleted crop lands. This can result in a relatively large collection area for a small amount of biomass, as only a fraction of the land in proximity to the plant is considered to be marginal and available for energy crop cultivation.

With the exception of a few ideal locations, it is assumed that 4,000 dry tons per day (tpd) of biomass will be the maximum economically feasible supply available for the majority of biomass energy conversion facilities in the United States, should widespread deployment occur. This assumption is congruent with a number of regional sites which can support this production rate from a collection radius of 30 to 50 miles based on the use of only marginal lands and without causing land use changes [6, 15]. This feed rate would result in 10.2-ton biomass delivery trucks arriving roughly every 10 minutes at a central collection facility.

2.4.5. Plant Size

The production capacity of the CTL/CBTL/BTL plant was based on three factors: (1) FT reactors are generally sized at 5,000 bpd, (2) at 50,000 bpd the effects of economies of scale taper off, making this the smallest plant size at which adding capacity will not significantly decrease capital requirements on a "dollars per daily barrel" basis, and (3) in the case of CBTL plants being fed biomass, the maximum biomass feed rate is 4,000 dry tpd.

Therefore, 50,000 bpd was viewed as the preferred CTL/CBTL/BTL plant size, unless this required a higher biomass feed rate than 4,000 dry tpd. In these cases, the plant size was decreased in 2,500 bpd increments until the required biomass feed rate was equal to or less than 4,000 dry tpd. As shown in Section 2.5, this resulted in "biomass-only" plants sized at 5,000 bpd production capacity and "co-gasification of coal and biomass plants"

that are fed 30% biomass (by weight), sized at 30,000 bpd. These reduced plant sizes do have an effect on the capital requirement for building the CTL/CBTL/BTL plant, resulting in significant increases in cost for the 5,000 bpd plant cost. This has a number of ramifications, including that the plant siting of BTL plants or CBTL plants which utilize large percentages of biomass may only be viable in special locations where larger biomass feed rates are available. Such opportunities are thought to be somewhat limited, however, and their assessment is beyond the scope of this study.

2.5. Case Descriptions

Table 2-1 describes the cases covered in this study. Three "Plant Types" are evaluated – CBTL, CTL, and BTL – at capacities up to 50,000 bpd. The plant capacity for plants fed with biomass was based upon a biomass feed rate of 4,000 dry tpd, which correlates to a biomass collection radius of thirty to fifty miles. This feed rate can be supported in a wide range of locations around the country without displacing food crops [15]. Feeds of fifteen and thirty percent biomass (by weight) were evaluated for the CBTL cases.

Although both the CBTL and BTL technologies can gasify a variety of biomass types, this study placed a focus on the use of switchgrass, which is an energy crop that can be grown on degraded or abandoned land. The coal feedstock chosen for the CBTL and CTL cases was Illinois #6, a high-sulfur, bituminous coal.

Table 2-1. CBTL Plant Configurations for this Study

Case	Plant Type	Capacity (BPD)	Biomass %	Biomass Type	Specification
1	CTL	50,000	n/a	n/a	No CCS
2	CTL	50,000	n/a	n/a	CCS
3	CTL	50,000	n/a	n/a	CCS + ATR
4	CBTL	50,000	8%	Switchgrass	CCS
5	CBTL	50,000	15%	Switchgrass	CCS
6	CBTL	50,000	15%	Switchgrass	CCS + ATR
7	CBTL	30,000*	30%	Switchgrass	CCS
8	CBTL	30,000*	30%	Switchgrass	CCS + ATR
9	BTL	5,000*	100%	Switchgrass	No CCS
10	BTL	5,000*	100%	Switchgrass	CCS
11	BTL	5,000*	100%	Switchgrass	CCS + ATR

*Plant capacity reduced from 50,000 BPD due to a scenario in which there is limited availability of biomass (4,000 dry tons per day).

2.6. Summary

CTL/CBTL/BTL is a process that combines several *existing* and *proven* technologies for the production of diesel fuel from both biomass and coal. The process produces a nearly pure stream of CO_2 which is ready for sequestration, resulting in a very low incremental cost to produce a fuel which has a life cycle GHG emissions profile which is less than that of petroleum-derived fuels, as will be detailed in Chapter 3 (GHG emissions levels) and Chapter 4 (economics).

CTL/CBTL/BTL is unique in that it *uses coal to make the use of biomass economically viable* by enabling large scale operation (driving down costs), preventing plant downtimes if biomass is not available, and providing a cheap, energy dense feedstock to lower costs. This makes it a very attractive solution for producing affordable, low-carbon diesel fuel from domestic resources, thereby enhancing energy supply security, promoting economic sustainability and addressing climate change issues associated with the transportation sector.

3. Global Climate Change and Fuel GHG Emission Profiles

This chapter discusses the life cycle GHG footprint of CTL/CBTL/BTL produced diesel fuel, compares these fuels to petroleum-derived diesel, and provides a detailed look at GHG emissions from the CTL/CBTL/BTL process. *CTL/CBTL/BTL offers a near-term opportunity for reducing GHG emissions* in the transportation sector by producing diesel fuel with a life cycle GHG intensity substantially (5% to 358%) below that of petroleum-derived fuels. As will be shown in Section 3.3, *reductions of 5-12% can be achieved in CTL plants and up to a 75% reduction is achievable when co-gasifying coal with commercially demonstrated levels of biomass (30% biomass by weight) in CBTL plants.* Using only biomass as a feedstock will produce a fuel with an even lower GHG footprint - up to 358% below that of petroleum - but this option is not likely to be economically viable, as is discussed in Chapters 4 and 5.

3.1. Alternative Fuel Criteria for Acceptability

A confluence of the desire for both (a) energy security and (b) GHG emissions reductions in the transportation sector has prompted a search for fuels which might supplant petroleum-derived fuels. One criterion used to inform policy and regulatory decisions regarding these fuels are the life cycle GHG emissions associated with these fuels, starting with the acquisition of raw materials from the earth (crude oil, coal, biomass, etc.) all the way through the use of the fuel in a vehicle.

Evaluating transportation fuels on a life cycle GHG emission basis is non-trivial, however, and care must be taken lest viable fuel options be needlessly precluded from development. For example, in April of 2007 the U.S. Environmental Protection Agency (EPA) published the results of a life cycle GHG analysis of 14 alternative transportation fuels in a 3-page fact sheet. One of the findings was that fuel produced by a CTL plant equipped with CCS had GHG emissions which were 3.7% *greater* than petroleum-derived diesel fuel, using fuel produced in the year 2017 as a basis of comparison (EPA 2007). These preliminary findings from the EPA led lawmakers to insert language into the Energy and Independence & Security Act (EISA) of 2007 to preclude the use of fuels with a higher GHG footprint than those produced from petroleum, effectively discouraging domestic CTL development.

This study clearly demonstrates that the use of the EPA feasibility study resulted in a misguided characterization of the life cycle GHG benefits of CTL with CCS. Through detailed analysis, NETL has found FT diesel fuel from CTL with CCS to have life cycle GHG emissions which are *9% to 15% below* that of petroleum-derived diesel, when a petroleum base year of 2017 is assumed (as in the EPA study). Figure 3-1 depicts the original April 2007 EPA bar chart and contrasts these findings with the 3.7% increase originally reported by the EPA.

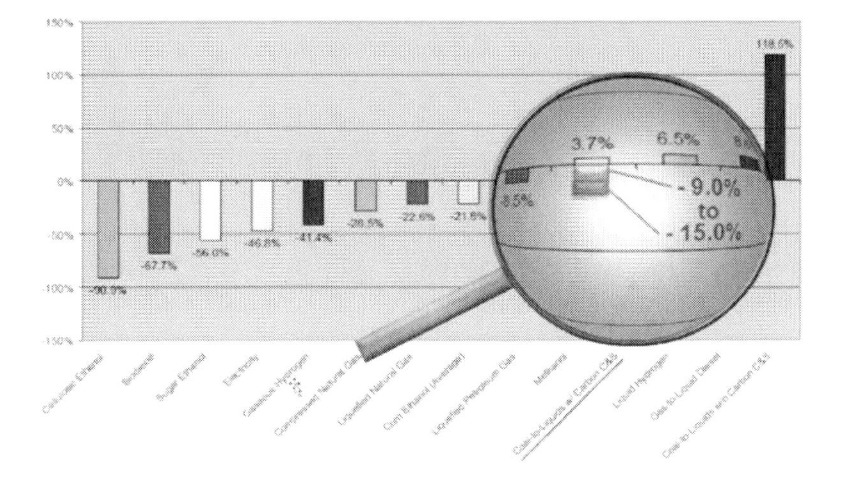

Figure 3-1. Percent Change in Emissions Using Non-EISA Petroleum Baseline.

3.1.1. Energy Independence and Security (EISA) Act of 2007 Provisions

EISA 2007, enacted December 19, 2007, contains two key provisions which pertain to the life cycle GHG emissions of alternative transportation fuels: the definition of a base year of comparison for GHG emissions and the requirement that alternative fuels procured by the government must meet a certain level of GHG emissions

3.1.1.1. 2005 Petroleum Baseline

The CTL/CBTL/BTL diesel fuels evaluated in this study are compared to the average emissions profile of petroleum-derived diesel, based on the U.S. national average in 2005. This "petroleum baseline" was used in order to be consistent with language in EISA 2007, which established the year 2005 as the basis of comparison for certain alternative fuels.[19] The sole exception to this is the result reported in Figure 3-1, where CTL with CCS is reported as a 9% to 15% reduction in life cycle GHG emissions compared to a petroleum base year of 2017, whereas it is 5% to 12% below the 2005 baseline. The 2017 baseline was used in this figure for consistency with the April 2007 EPA document and will not be referred to again.

As the U.S. EPA Administrator has not determined a 2005 petroleum baseline as of December 2008; the 2005 petroleum baseline reported in the NETL report entitled "Development of Baseline Data and Analysis of Life Cycle Greenhouse Gas Emissions of Petroleum-Based Fuels" is used as a basis of comparison in this study [16]. This study is discussed in greater detail in Appendix B.

3.1.1.2. Life Cycle Emissions Comparison to Petroleum-Derived Diesel

As mentioned above, when faced with a direct result of the EPA preliminary finding, lawmakers inserted language into Title V, Subtitle C, Sec. 526 of EISA 2007 which precluded federal agencies from entering "into a contract for procurement of an alternative or synthetic fuel...unless the contract specifies that the lifecycle greenhouse gas emissions associated with the production and combustion of the fuel supplied under the contract must...be less than or equal to such emissions from the equivalent conventional fuel produced from conventional petroleum sources." Based on this criteria and the EISA language which sets 2005 for the base year for comparison, CTL with CCS qualifies for federal procurement, as is discussed in Section 3.3.

3.2. GHG Emissions from Petroluem-Derived Diesel Fuel

The production and delivery of transportation fuels has been widely studied in the United States. Over the past 10 years, the increasing emphasis on GHG emissions in the United States and abroad has resulted in a number of well-documented and cited reports on the life cycle emissions of petroleum-derived diesel fuel. Figure 3-2 describes the "Well-To-Tank" (WTT) GHG emissions for petroleum diesel, as reported in a number of these studies.[20] As shown, the results vary widely across different crude sources, base years, and modeling assumptions, ranging from 11.8 kg CO_2E/million British thermal units (mmBtu) to 37.5 kg CO_2E/mmBtu – a three-fold differential between the minimum and maximum values.

In November of 2008, NETL released a detailed study which found the WTT GHG emissions profile of petroleum-derived diesel fuel to be 18.3 kg CO_2E/mmBtu lower heating value (LHV) of diesel fuel dispensed, based on the average U.S. transportation fuel sold or distributed in 2005. This result, shown as the green bar in Figure 3-2, has half the emissions of Venezuela very heavy crude, as estimated by McCann in 1999. This has the clear ramification that by displacing the marginal supply of heavy crude imports with FT fuels which have GHG emissions equal to or below the petroleum baseline, the overall GHG emissions from the transportation sector can be dramatically reduced.

When the "Well-to-Wheels" (WTW) life cycle emissions are considered and vehicle operation is included in the emissions profile, the total life cycle GHG emissions are 95.0 kg CO_2 E/mmBtu LHV of fuel consumed (or 7.3 kilograms per gallon of diesel fuel consumed) [16].[21] As described in Figure 3-3, the bulk of these emissions (81%) are associated with vehicle operation during which carbon in the fuel is combusted and converted to CO_2. Since the capture and disposal of these combustion emissions is not viewed as viable with conventional vehicle technology, the maximum GHG emissions reduction possible achievable by making changes to the existing petroleum-based diesel production chain is 19% – or 19% below the 2005 average petroleum baseline – assuming all upstream emissions are eliminated.[22]

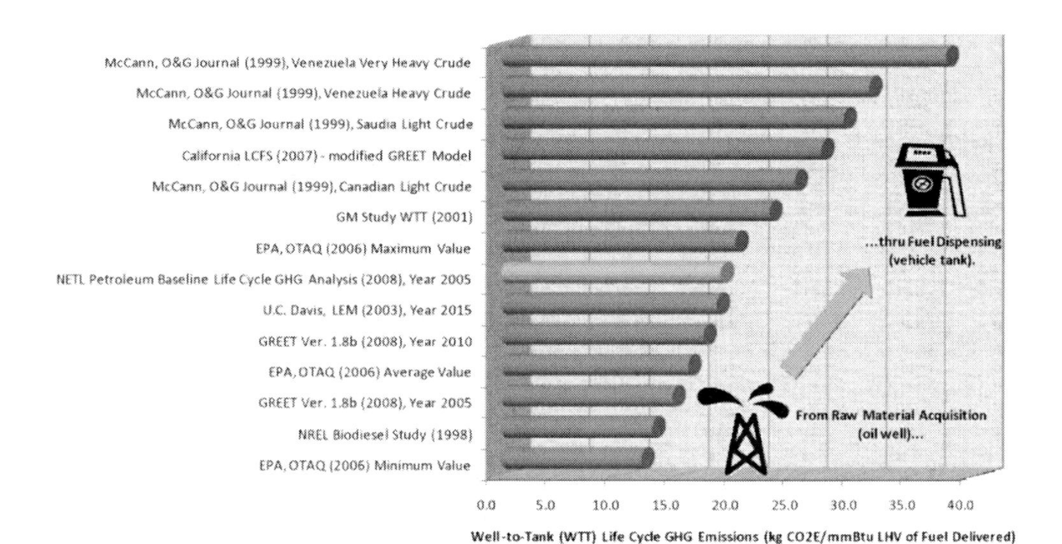

3.3. GHG Emissions from FT Diesel

CTL/CBTL/BTL can produce diesel which has a GHG emissions profile significantly below that of petroleumderived diesel. This represents a unique opportunity for GHG emissions reductions from the transportation sector in light of the limited reductions achievable from petroleum-derived diesel (19% theoretical maximum). This reduced GHG profile is achieved in two ways: carbon sequestration and the use of biomass to supply all or part of the required feedstock energy. Carbon sequestration is utilized to reduce upstream emissions of the fuel to a level below the upstream emissions of petroleum-diesel. The use of biomass as a feedstock offsets a portion of the carbon released during combustion, reducing the "vehicle operation" emissions. Biomass cultivation can also result in the sequestration of carbon in low-grade soils, further reducing GHG emissions, although credit for this GHG offset was not taken into account by this study.[23]

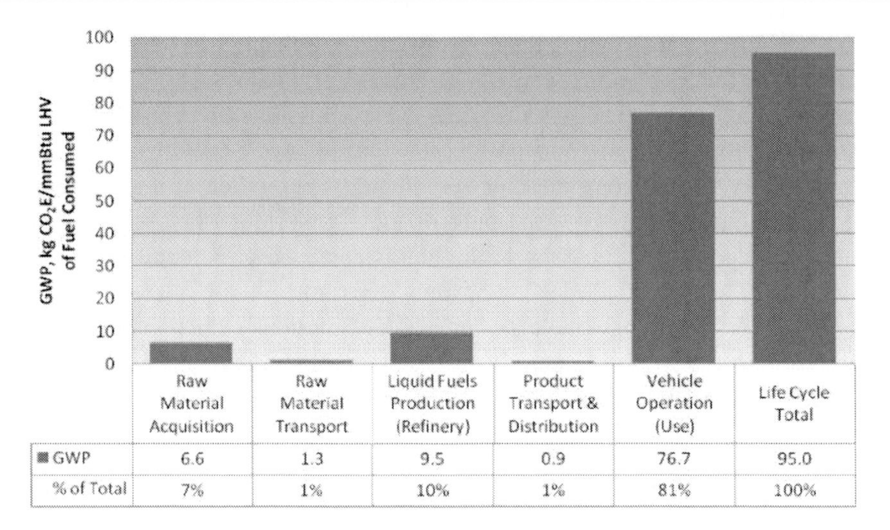

Figure 3-3. Breakdown of Life Cycle GHG Emissions from Petroleum-Derived Diesel.

When carbon sequestration and the use of biomass are combined, the diesel produced has a fraction of the GHG emissions of petroleum-derived fuels and can even be "GHG negative", where more carbon is removed from the atmosphere than is emitted. The improvement is even more pronounced for BTL plants in which only biomass is used as a feedstock because some of the CO_2 removed from the atmosphere during photosynthesis is then sequestered. This benefit is offset by limits of available biomass in the nation, therefore reducing the total amount of fuel produced and the total impact of the lower-carbon profile fuel, as is described in Chapter 5.

3.3.1. Emissions Profiles FT Diesel

The life cycle GHG emissions of the FT diesel produced in Cases 1 to 11 were calculated using the methodology described in Appendix B and then compared to the petroleum baseline. Of the ten cases evaluated, only one – CTL without CCS – produced a fuel with a higher GHG emissions profile than the petroleum baseline. As detailed in Table 3-2, the remaining cases range from 5% below that of petroleum (CTL with CCS) to 358% below the baseline (BTL with CCS+ATR) with emissions reductions increasing steadily with increased biomass percentages in the feed, and likewise, as an ATR is added to the CCS configurations.

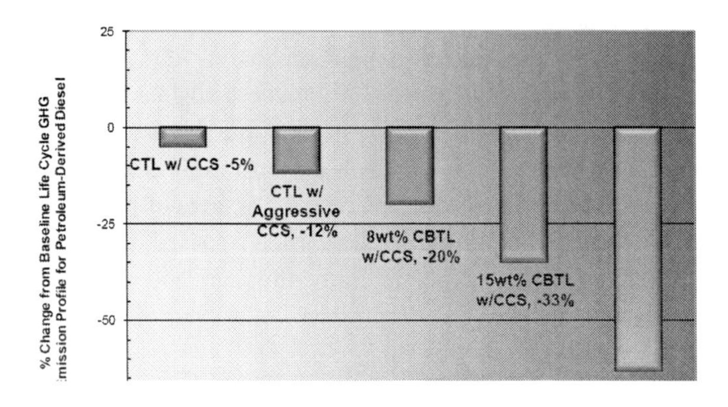

Figure 3-4. GHG Emissions of CBTL Plants Compared to Petroleum-Derived Diesel Fuel.

Table 3-2. GHG Emissions of CBTL Plants Compared to Petroleum-Derived Diesel Fuel

Case	1	2	3	4	5	6	7	8	9	10	11
Description	CTL	CTL	CTL	CBTL	CBTL	CBTL	CBTL	CBTL	BTL	BTL	BTL
CCS	None	Simple	ATR	Simple	Simple	Simple	ATR	ATR	None	Simple	ATR
Biomass %	n/a	n/a	n/a	8wt%	15wt%	30wt%	15wt%	30wt%	100%	100%	100%
WTW GHG Emissions (kg CO_2eq/mmBtu)	235	90.2	83.7	76.0	63.4	35.1	55.3	23.8	-8.8	-210.0	-245.0
% Change from Petroleum	+147%	-5%	-12%	-20%	-33%	-63%	-42%	-75%	-9.2%	-321%	-358%

The five cases described in Figure 3-4 represent the salient CTL/CBTL/BTL plant configurations. The first key finding is that the addition of CCS to a CTL plant can result in fuels which have 5% to 12% less GHG emissions than petroleum-derived diesel. Furthermore, this can be achieved at a low incremental cost, as will be discussed further below. This shows that FT diesel can provide a pathway to domestic energy security while improving GHG emissions from the transportation sector.

The second finding is that co-gasifying a modest amount of biomass – 8% by weight – reduces the GHG emissions profile of the fuel to 20% below that of petroleum-derived diesel. Co-gasification of 15% and 30% (by weight) biomass results in emissions reductions of 33% and 63%, respectively.[24] Adding an ATR to the cases that co-convert coal and biomass will reduce emissions further but is not an economically preferred option, as will be shown in Chapter 4.

3.3.2. Carbon Flows and GHG Emissions

Figures 3-5 and 3-6 illustrate the carbon flows in two representative cases: CTL with CCS and 15wt% CBTL with CCS. In each figure, the blue lines represent carbon flows into and out of the process. The width of these lines represents the relative carbon flow rate, listed in short tons of carbon equivalents per day. The red lines in the field are GHG emissions from other processes such as coal mining, biomass cultivation, and fuels transportation, and again, the line width represents the relative amount of emissions, listed in short tons of carbon equivalents per day.

Figure 3-5 depicts the carbon flows and other net GHG emissions for Case 2: CTL with CCS. Carbon – in the form of coal – is removed from the ground (shown as the horizontal dotted line near the bottom of the figure) and converted into liquid fuels via the FT Process. The 13,522 tons/day of carbon supplied by coal is disposed of as follows: a) 7,267 tons/day is captured at the plant and sequestered as CO_2, b) 720 tons/day is vented from the plant as CO_2, c) 135 tons/day is discharged from the plant in the form of char, d) 3,770 tons/day is transferred to the diesel fuel and released as CO_2 during its use, and e) 1,630 tons/day is transferred to the naphtha co-product, which displaces petroleum-derived naphtha.[25]

There are three sources of GHG emissions which are not directly associated with the carbon in the process: (1) coal mining and transportation: 495 tons/day; (2) transportation of the finished FT diesel fuel: 11 tons/day; and (3) a displacement credit taken for the upstream GHG emissions associated with the production of naphtha from petroleum: -358 tons/day. These emissions contribute to the total life cycle GHG emissions of the FT diesel, or, in the case of the naphtha displacement credit, reduce the life cycle emissions based on emissions which do not occur elsewhere.

The horizontal dotted line towards the top of Figure 3-5 can be considered to be the atmosphere, and by summing the carbon (or carbon equivalent) flows which are above that line, the net GHG emissions to the atmosphere can be calculated. The net life cycle emissions for Case 2 are 4,638 tons of carbon per day. Dividing this by the heating value of the finished FT diesel product (and converting from tons of carbon to tons of CO_2) gives the WTW emissions of the fuel, which are found to be 90 kg CO_2eq/mmBtu (LHV) of fuel. The WTT emissions are calculated by subtracting the 3,770 tons/day of Combustion Emissions from the net emissions (4,638 tons/day), then dividing by the FT diesel fuel energy content, yielding WTT emissions of 17 kg CO_2eq/ mmBtu (LHV) of fuel.

Figure 3-6 is similar to Figure 3-5, but depicts a CBTL plant, fed with both coal and biomass. In this case, 12,062 tons/day of the carbon supplied to the process is coming from coal (and from under the ground), while 1,412 tons/day of the carbon is coming from biomass. This 1,412 tons/day was originally in the atmosphere in the form of CO_2 before it was converted into carbon via photosynthesis, *therefore this carbon is considered a negative carbon flow when summing the net emissions to the atmosphere.* The fate of the 13,474 tons/day of carbon entering the 15wt% CBTL plant is similar to that entering the CTL plant: a) 7,267 tons/day is captured at the plant and sequestered as CO_2, b) 671 tons/day is vented from the plant as CO_2, c) 135 tons/day is discharged from the plant in the form of char, d) 3,774 tons/day is transferred to the diesel fuel and released as CO_2 during its use, and e) 1,627 tons/day is transferred to the naphtha co-product, which displaces petroleum-derived naphtha.

Note that while the liquid fuels output is the same in both cases, slightly less CO_2 is vented to the atmosphere in the 15wt% CBTL case. This is due to a slight increase in thermal plant efficiency for the 15wt% CBTL case, which results in greater levels of carbon conversion by the process. This illustrates how future improvements or modifications to the CBTL plant can reduce GHG emissions from the plant, thereby improving the GHG emissions profile of the FT diesel produced.[26]

The WTW and WTT emissions for Case 5 are calculated in a manner similar to that described above for Case 2, except that as noted, the carbon which originally came out of the atmosphere to become the biomass is counted as a negative carbon flow, resulting in total net GHG emissions to the atmosphere of 3,261 tons/day. Thought of another way, the removal of this CO_2 from the atmosphere serves to offset CO_2 released into the atmosphere during the combustion of the FT diesel fuel, and is therefore a portion not counted toward the total carbon emitted and therefore reducing the emissions profile of the fuel.

3.4. Summary

CTL/CBTL/BTL offers a near-term opportunity for reducing GHG emissions in the transportation sector by producing a low GHG intensity diesel fuel which is a drop-in replacement for petroleum-derived diesel. In CTL plants, a fuel can be produced which has 5% and 12% less life cycle GHG emissions than petroleum-derived diesel, using carbon sequestration, and sequestration coupled with aggressive capture, respectively. Therefore, CTL with CCS clearly meets the EISA 2007 criteria of producing a fuel with less life cycle GHG emissions than petroleum-derived diesel and federal agencies will be able to procure this fuel.

Furthermore, by co-gasifying a modest amount of biomass – 8% by weight –the GHG emissions profile of the fuel is reduced to 20% below that of petroleum-derived diesel. Co-gasification of 15% and 30% (by weight) biomass results in emissions reductions of 33% to 63%, respectively. Additional reductions in GHG emissions can also be achieved through the use of aggressive CCS.

Using only biomass as a feedstock will produce a fuel with an even lower GHG footprint – up to 358% below that of petroleum – but this option is not likely to be economically viable, as is discussed in Chapters 4 and 5.

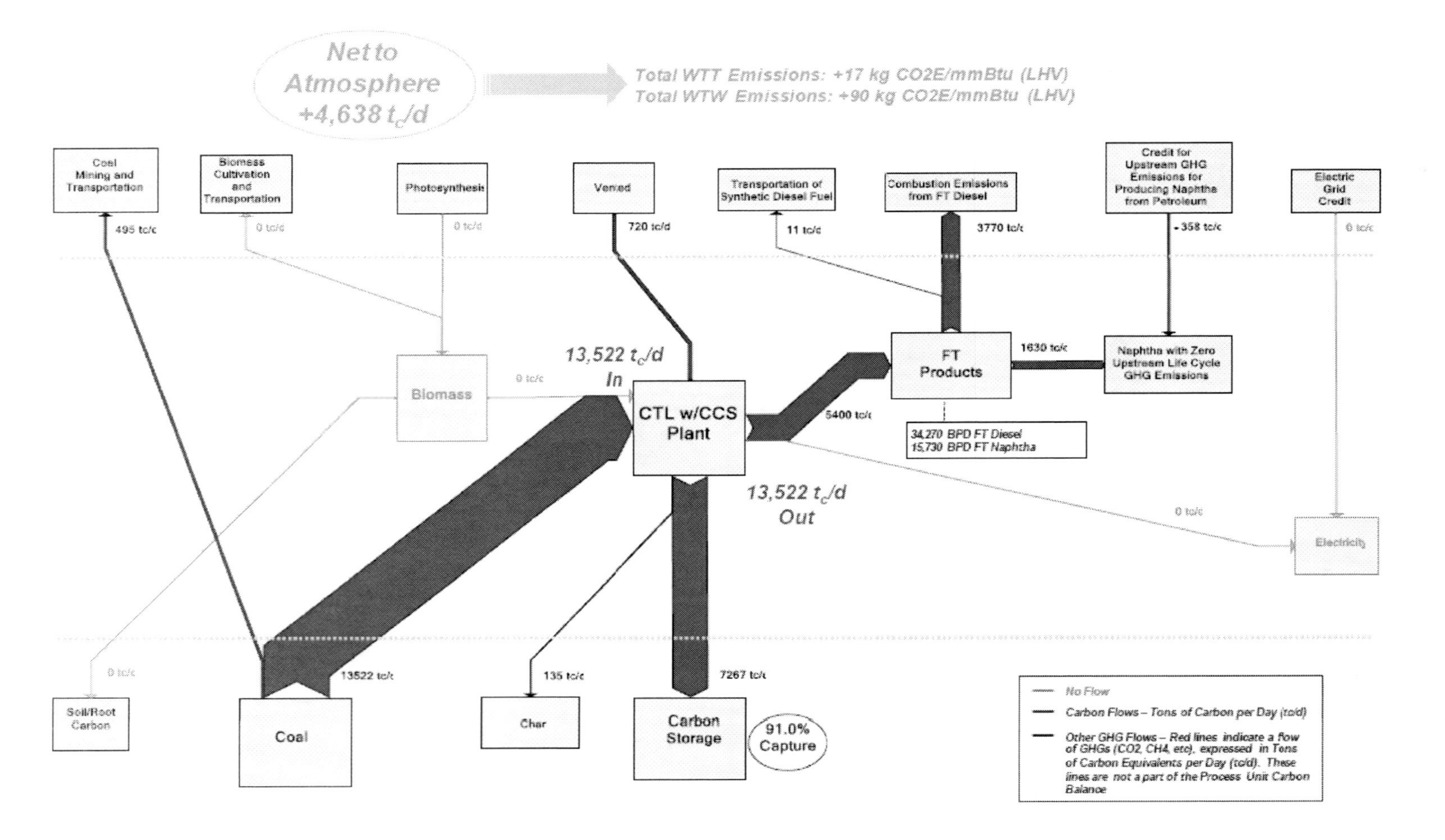

Figure 3-5. Carbon Flows and Life Cycle GHG Emissions from CTL with CCS.

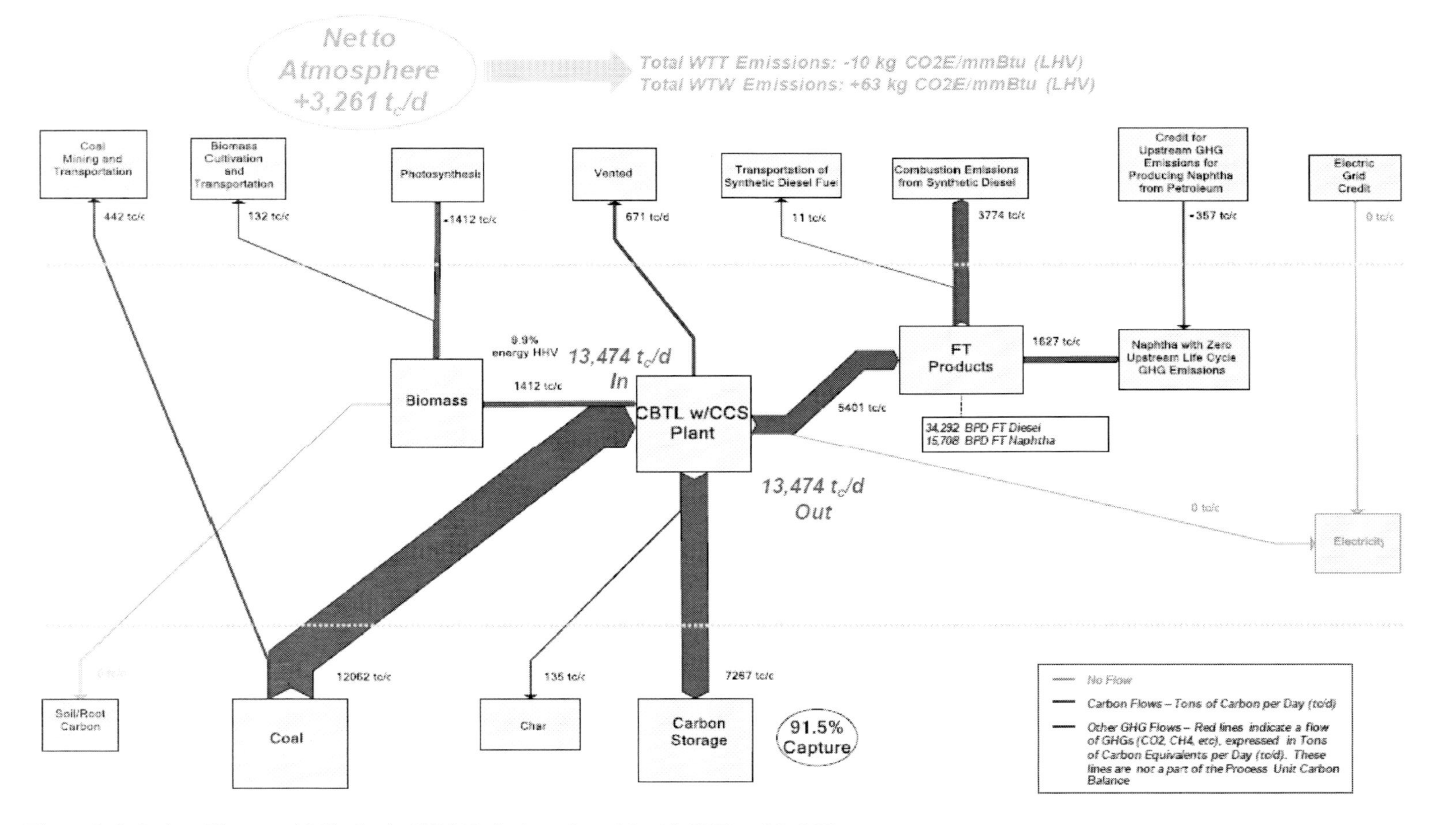

Figure 3-6. Carbon Flows and Life Cycle GHG Emissions from 15wt% CBTL with CCS.

4. Economic Feasibility of CBTL Processes

As seen in Chapter 3, CTL/CBTL/BTL plant configurations exist which can produce diesel fuel with low or even net negative life cycle GHG emissions. However, these plant configurations are not always the economically preferred CTL/CBTL/BTL plant configuration.

This chapter details an analysis which derives the Required Selling Price (RSP) of the FT diesel fuels produced in order to determine the economic feasibility and relative competitiveness of the different plant options. A sensitivity analysis was performed to determine how carbon control regulations such as an emissions trading scheme for transportation fuels would affect the price of both petroleum-derived diesel and FT diesel from the different plants.

The key findings of these analyses were:

1) CTL plants equipped with CCS are competitive at crude oil prices as low as $86/bbl and have less life cycle GHG emissions than petroleum-derived diesel. These plants become more economically competitive as carbon prices increase.
2) The incremental cost of adding simple CCS is very low (7 cents per gallon) because CO_2 capture is an inherent part of the FT process. This becomes the economically preferred option at carbon prices above $5/_{mtCO2eq}$.[27]
3) BTL systems are hindered by limited biomass availability which affects the maximum plant size, thereby limiting potential economies of scale. This, combined with relatively high biomass costs results in FT diesel prices which are double that of other configurations: $6.45 to $6.96/gal compared to $2.56 to $2.82/gal for CTL and 15wt% CBTL systems equipped with CCS.

The conclusion reached based on these findings was that both the CTL with CCS and the 8wt% to 15wt% CBTL with CCS configurations may offer the most pragmatic solutions to the nation's energy strategy dilemma: GHG emission reductions which are significant (5% to 33% below the petroleum baseline) at diesel RSPs that are only half as much as the BTL options ($2.56 to $2.82/gal compared to $6.45 to $6.96/gal for BTL). These options are economically feasible when crude oil prices are $86 to $95/bbl.

4.1. A Hypothetical Carbon Control Regulation

In addition to their positive impact on our nation's energy security, a key motivation for the development of CBTL and BTL plants is the prospect of a carbon control regulation. Accordingly, to assess the potential impact of such a regulation on the economic feasibility of CTL/CBTL/BTL plants, a hypothetical regulatory structure was assumed under which a certain cost ($/mt$CO_2$eq) would be incurred for a transportation fuel's life-cycle GHG (raw material extraction thru fuel use). If a fuel achieves *negative* life-cycle GHG emissions, a credit would be applied at the same rate.

For this analysis, it was assumed that the full cost (or credit) associated with the life-cycle GHG emissions of each diesel fuel product was embedded into its spot price. This single adjustment accounts for all the life-cycle GHG emission costs, which are actually incrementally accumulated at different points along the production chain, both upstream and downstream.

4.2. Economic Assessment Methodology

A key measure of the economic feasibility of a CTL/CBTL/BTL project is the RSP.[28] This is the minimum price at which its FT diesel fuel product must be sold in order to: a) offset its operating costs (including the cost of GHG emissions), b) service its debt, and c) provide the expected rate of return to its equity investors. If the market price of diesel fuel is equal to or above the RSP, the CTL/CBTL/BTL project is considered economically feasible.

Two key considerations must be taken into account when comparing the RSP to the market price of diesel, however, namely (1) the difference in energy content between petroleum-derived diesel (which the FT diesel is expected to complete with, in the market) and, (2) whether a GHG regulatory policy exists that pertains to transportation fuels. The latter point is especially important given the cost premiums associated with producing biomass-derived fuels: the price of carbon emissions can be the deciding factor as to whether these low GHG intensity fuels are feasible.

To address these considerations, the RSP of all FT diesel products were normalized to a $ per gallon petroleum-diesel equivalent basis, as described below in Section 4.2.2 , and the economic viability of all CTL/CBTL/BTL plants was evaluated over a range of carbon prices, as described in Section 4.3. In these scenarios, the petroleum-derived diesel price was defined as a function of both crude oil price and carbon emissions price. This has the secondary benefit of allowing CTL/CBTL/BTL plant feasibility to be evaluated in terms of a constant crude oil price (COP), a figure which is often more familiar to the public, while varying the carbon price, as is described below.

The remainder of this section describes the methodology used for the: (1) calculation of the RSP, (2) normalization of FT diesel to petroleum-derived diesel equivalents, and (3) relationship of petroleum-derived diesel to COP and GHG emission value.

4.2.1. Required Selling Price

The RSP is the minimum price at which the FT diesel produced by a CTL/CBTL/BTL plant must be sold in order to: a) offset its operating costs (including the cost of GHG emissions), b) service its debt, and c) provide the expected rate of return to its equity investors.

A discounted cash flow analysis was conducted to calculate the diesel RSP for a variety of CTL/CBTL/BTL plant configurations. In this study, it was assumed that CTL/CBTL/BTL projects must achieve a 20% Internal Rate of Return on Equity (IRROE) to be economically feasible. The project finance structure was selected to reflect a hypothetical mid-term future in which regulatory risk has been eliminated by the passage of a carbon regulation, and technical risk has been partially mitigated by the demonstration of two or three commercial-scale CTL/CBTL/BTL plants. In addition, the project finance structure is assumed to benefit from a moderate government incentive, such as a loan guarantee. No other incentives are assumed for any of the CTL/CBTL/BTL cases. For more details on the economic analysis methodology and its key assumptions, see Appendix A.

4.2.2. Normalization of FT Diesel to Petroleum-Derived Diesel

To properly compare the RSP of the diesel produced by a CTL/CBTL/BTL plant to the price of petroleum-derived diesel produced by a refinery, the two prices must have an energy equivalent basis. In this study, the LHV of FT diesel is 118,905 Btu/gal, which is 9% lower than the LHV assumed for petroleum-derived diesel (131,229 Btu/gal). Therefore, to express the FT diesel RSP in terms of "dollars per gallon petroleum-diesel equivalent" the Equation 4-1 can be used.

Equation 4-1. FT Diesel RSP, Normalized to Petroleum-Derived Diesel

$$\text{RSP (\$/gal petroleum-diesel equivalent)} = [\ 131{,}229 / 118{,}905\ ] \times \text{RSP (\$/gal FT diesel)} \qquad (4\text{-}1)$$

4.2.3. Crude Oil and Petroleum-Derived Diesel Price Correlation

To assess whether a CTL/CBTL/BTL project will be economically feasible over its operating life, one must compare its diesel RSP to the expected future market price of ultra-low-sulfur diesel (ULSD) (on an energy equivalent basis). This analysis assumes that the market price will be equal to the price obtained by U.S. refineries for their petroleum-derived, ULSD product.

Accordingly, it is necessary to calculate the price of petroleum-derived, ULSD fuel as a function of two key variables that are of interest in this study: the crude oil price (COP) and the GHG emission value. Equation 4-2, below, is used for this calculation.

Equation 4-2. Price of Petroleum-Derived Diesel as a Function of Crude Oil and GHG Emission Value

$$\text{PDDP} = [\ 1.25 \times (\text{COP \$/bbl}) + (0.524\ \text{mtCO}_2\text{eq/bbl}) \times (\text{GHGEV \$/mtCO}_2\text{E})\ ] / (42\ \text{gal/bbl}) \qquad (4\text{-}2)$$

where:

PDDP = Petroleum-derived diesel price (ultra-low-sulfur), \$/gal

COP = West Texas Intermediate (Cushing, OK) crude oil spot price, \$/bbl GHGEV = Greenhouse gas emission value, \$/mtCO_2E

To account for the "refiner's margin", the COP is multiplied by the historical ratio of the ULSD spot price (New York Harbor) to the crude oil spot price (West Texas Intermediate, Cushing, OK). For the period January 2002 through July 2008, this ratio was 1.25. For a detailed explanation of how this ratio was derived, see Appendix A, Section A.5.2.

To account for the cost of GHG emissions, the GHG Emission Value (GHGEV) is multiplied by the average life-cycle GHG emissions of petroleum-derived diesel fuel sold or distributed in the U.S. in 2005. This value is 0.524 mtCO_2eq per bbl. For a detailed explanation of how this value was estimated, see Appendix B, Section B.3. Note that this single adjustment accounts for all the life-cycle GHG emission costs, which in reality may be incrementally accumulated at different points along the production chain, both upstream and downstream. The factor of 42 is applied to convert from barrels to gallons.

Using Equation 4-2, the price of petroleum-derived diesel as a function of GHG emission value for several crude oil prices ranging from \$60 to \$110/bbl, as described in Figure 4-1. The y-intercepts reflect the diesel price when there is not a carbon regulation (GHG emission value is zero). As the GHG emission value increases (x-axis), the diesel price also increases because the cost of the associated life-cycle GHG emissions is embedded into it.

Using the above equations, one can calculate combinations of GHG emission values and crude oil prices that will result in a petroleum-derived diesel price that is equivalent to the FT diesel RSP. For example, the following equation can be used to calculate the "equivalent" crude oil price (\$/bbl) that results in parity between the petroleum-derived diesel price and the FT diesel RSP.

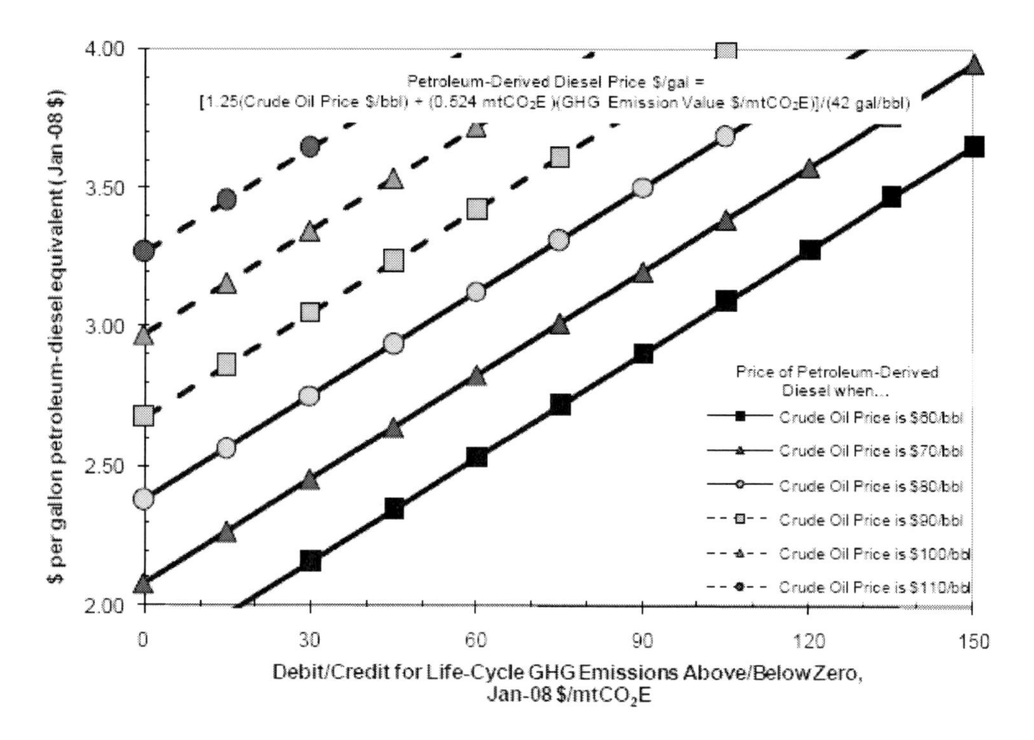

Figure 4-1. Effect of GHG Emission Value and Crude Oil Price On Petroleum-Derived Diesel Price.

These Equations, 4-1 through 4-3, were used throughout Sections 4.3 and 4.4 to evaluate and compare the economic feasibility of different CTL/CBTL/BTL configurations.

Equation 4-3: Equivalent Crude Oil Price where Diesel RSP Matches Petroleum-Derived Diesel Price

$$\text{Equivalent COP} = [(42\text{gal/bbl}) \times (\text{RSP\$/gal}) - (0.524\ \text{mtCO}_2\text{eq/bbl}) \times (\text{GHGEV \$/mtCO}_2\text{E})]/1.25 \tag{4-3}$$

where RSP is expressed in \$/gal petroleum-diesel equivalent

4.2. CBTL Capital Costs

The RSP, and therefore the economic feasibility of a CTL/CBTL/BTL project is sensitive to its capital cost. The methodology used to estimate capital cost is described in Appendix C, and detailed capital cost breakdowns for each CTL/CBTL/BTL case are provided in Appendix E. Appendix E also tabulates the estimated operating costs for each case.

Specific capital costs ($ per daily barrel (db) of fuel production capacity) are shown for selected cases in Figures 4-2 and 4-3, below. Figure 4-2 displays total overnight capital cost per daily barrel of production capacity. This is the sum of all capital cost elements excluding interest during construction and escalation during construction, expressed in Jan-2008 dollars and normalized to production capacity basis.[29] Figure 4-3 displays total as-spent capital cost, which is the sum of all capital cost expenditures over the construction period including interest and escalation during construction, expressed in mixed-year dollars per daily barrel of production capacity.

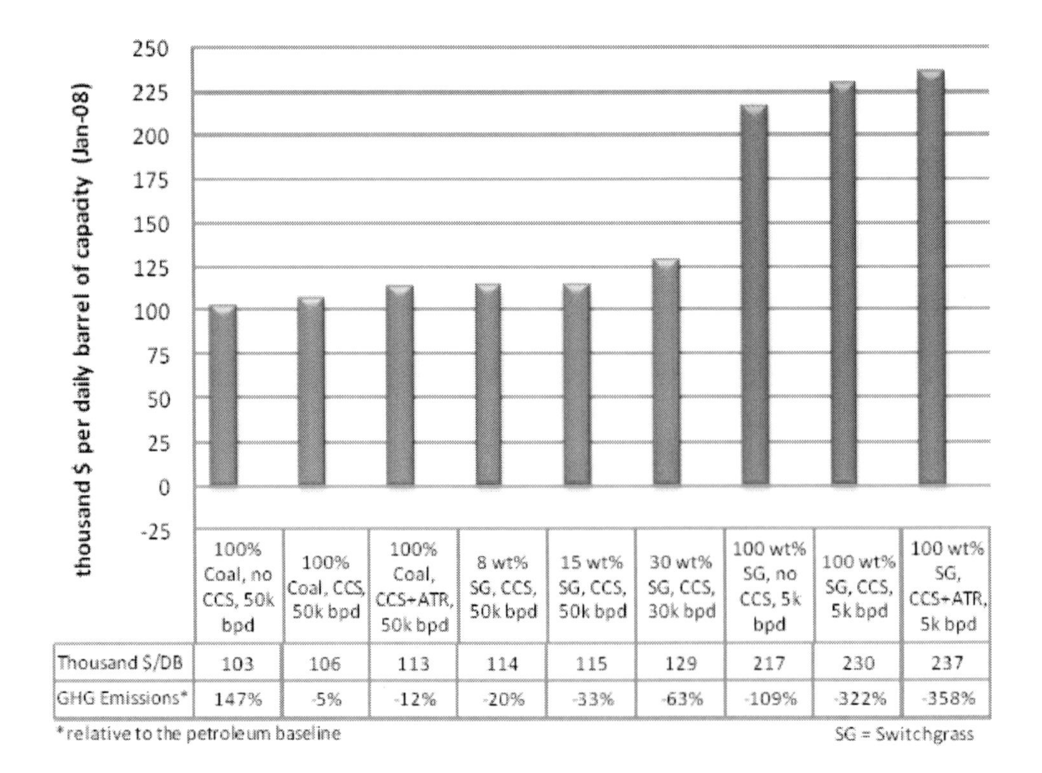

Figure 4-2. Total Overnight Capital Cost (excludes escalation & interest during construction).

The following conclusions can be drawn from the capital costs shown in Figures 4-2 and 4-3:

- Because CTL plants must capture ~91% of the CO_2 they generate as an inherent part of the FT process, adding the ability to compress and sequester the CO_2 is very inexpensive, increasing the as-spent capital cost by only $4,000 per daily barrel of capacity, or less than 4%. This small investment reduces the GHG emissions dramatically, from 147% above the petroleum baseline to 5% below it.

Unfortunately, achieving further reductions is more costly. For example, enhancing the capture rate to 96% reduces GHG emissions to a level that is 12% below the petroleum baseline but requires an incremental investment that is over twice as much: $11,000 per daily barrel of capacity, or a 10% increase. Adding the ability to handle, prepare and co-feed 8wt% or 15wt% biomass requires a similar increase in capital cost.

- The specific capital cost (dollars per daily barrel of production capacity) increases as the plant size decreases due to a loss of significant economies of scale. The specific capital costs of the BTL cases are twice as high primarily because comparable economies of scale cannot be attained. As discussed in Sections 2.4.4 and 2.4.5, the assumption that the biomass supply rate is limited to 4,000 dry tons per day constrains the capacity of the 30wt% CBTL plant to 30,000 bpd and the capacities of the BTL plants to only 5,000 bpd (compared to 50,000 bpd for the other plants).

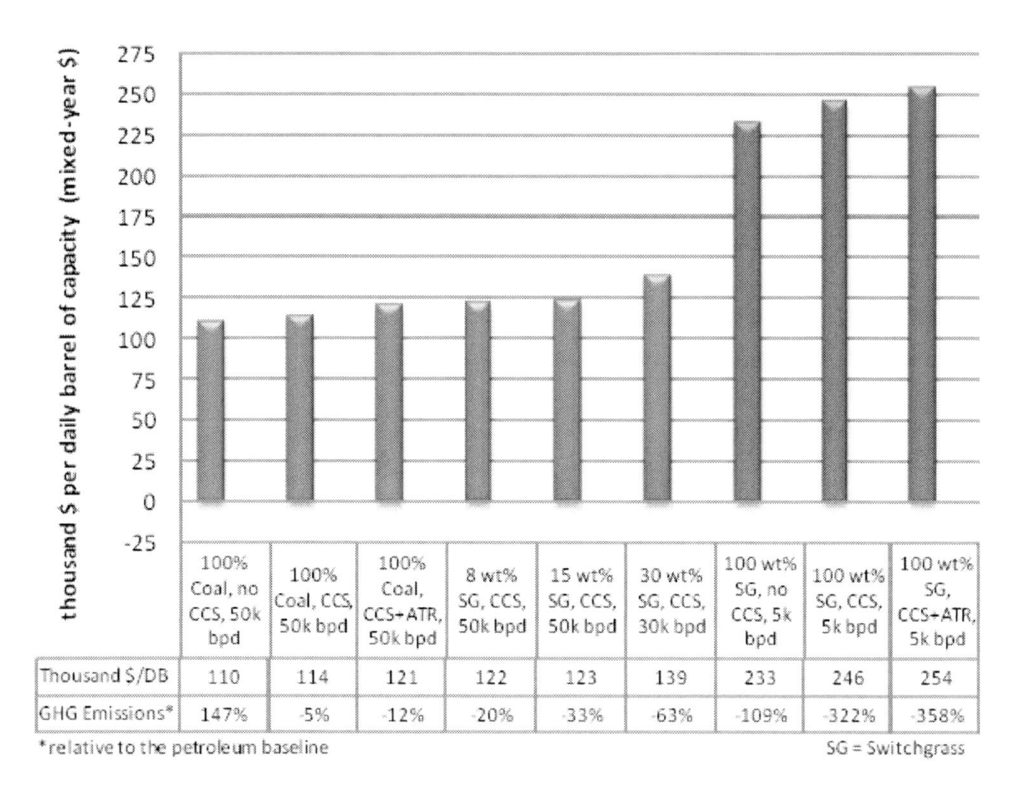

Figure 4-3. Total As-Spent Capital Cost (includes escalation & interest during construction).

Figure 4-4 shows a percentage breakdown of the various costs that comprise the total overnight capital cost for a CBTL plant. The gasification island is the largest component, comprising around one-third of the total cost. Note that the sum of all capital costs associated with CCS comprise only 6% of the total overnight capital cost, and much of this cost would be required by a CTL plant even when sequestration is not employed.

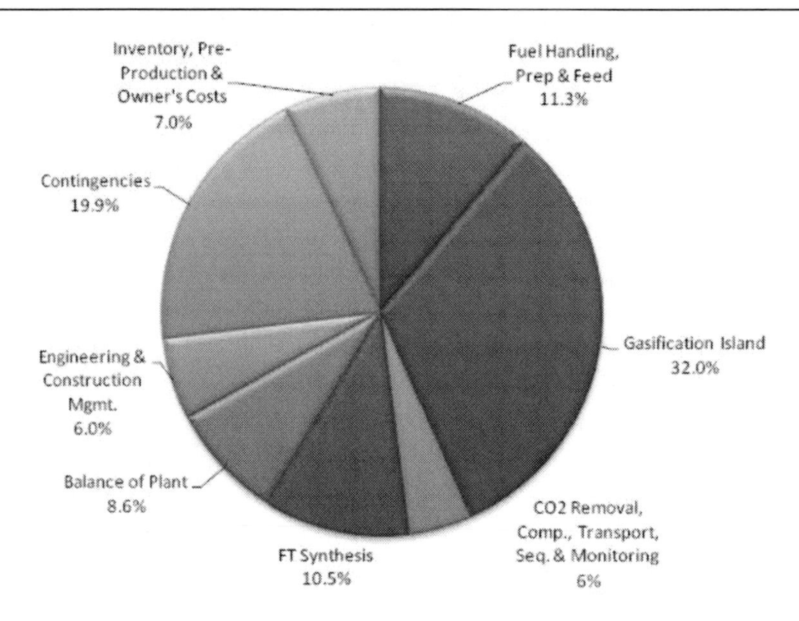

Figure 4-4. Distribution of Total Overnight Capital Costs for 15 wt% CBTL+CCS, 50k bpd.

4.3. CBTL Economic Feasibility

4.3.1. Biomass-Only Cases

In Figure 4-5, the RSP (y-axis) for the following three BTL cases (Cases 9 through 11) is graphed as a function of the GHG emission value (x-axis). Case 1 (CTL without CCS) is also shown on Figure 4-5 for the purpose of comparison.

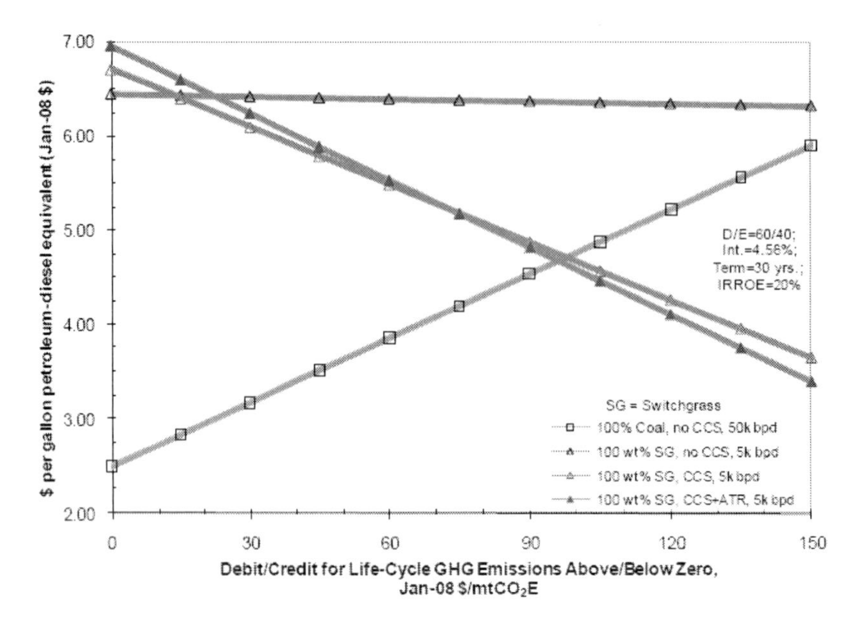

Figure 4-5. BTL Cases - Effect of GHG Emission Value on Diesel RSP.

The point at which each line crosses the y-axis in Figure 4-5 is the RSP when the GHG emission value is zero (\$0 /mtCO_2eq), reflective of a scenario in which there is no GHG regulation in effect. The following conclusions can be drawn from an examination of the y-intercepts on Figure 4-5:

- When GHG emissions have no value, CTL without CCS is economically viable when the diesel price is 2.49 dollars per gallon (\$/gal) or higher.
- When GHG emissions have no value, the BTL cases are not economically viable unless the diesel price is above \$6.45/gal.

Note that the slope of each line is proportional to the degree to which the plant's GHG emissions are above (positive slope) or below (negative slope) zero. A horizontal line would indicate the fuel has no net GHG emissions over its life cycle, and thus the RSP is completely insensitive to the GHG emission value. The CTL without CCS case is the only case in Figure 4-5 for which the diesel RSP increases with the GHG emission value. This is because it is the only case that has life-cycle GHG emissions that exceed zero. In contrast, the diesel RSP for each BTL case decreases as the value of GHG emissions increase, with the GHG emission value because they collect credits for having life-cycle GHG emissions below zero.

The intersections of the lines denote the GHG emission value at which two cases have the same diesel RSP. The following conclusions can be drawn from the intersections on Figure 4-5:

If GHG emission values are above \$14/mtCO_2eq, adding CCS to a BTL plant results in a lower diesel RSP than the other BTL options.

If GHG emission values are above \$73/mtCO_2eq, adding an ATR to increase the degree of carbon capture results in a lower diesel RSP than the other BTL options.

Considering only the cases displayed on Figure 4-5, the CTL case is the economically preferred alternative based on diesel selling price when GHG emissions are valued below \$96/mtCO_{2eq}; above this value, the BTL+CCS+ATR case is economically preferred. The other two BTL cases (BTL without CCS and BTL+CCS) are never economically preferred.

As will be shown in the following sections, when BTL is competed against CBTL options that reduce GHG emissions below the petroleum baseline, a BTL configuration is not the economically preferred alternative unless the GHG emission value exceeds \$138/mtCO_2eq. GHG emission values over \$138/mtCO_2eq are not expected to be economically or politically acceptable, since they would increase the cost of petroleum-derived diesel fuel by more than \$1.70 per gallon. Consequently, there is not a scenario foreseeable in which BTL would be economically feasible.

4.3.2. Coal-Only Cases

In Figure 4-6, lines are included for the CTL+CCS and CTL+CCS+ATR cases. The BTL+CCS+ATR is retained from Figure 4-5 because it remains the economically preferred option at high GHG emission values (the other BTL cases drop out). The scale of the y-axis has been decreased, compared to Figure 4-5, to magnify the range of interest.

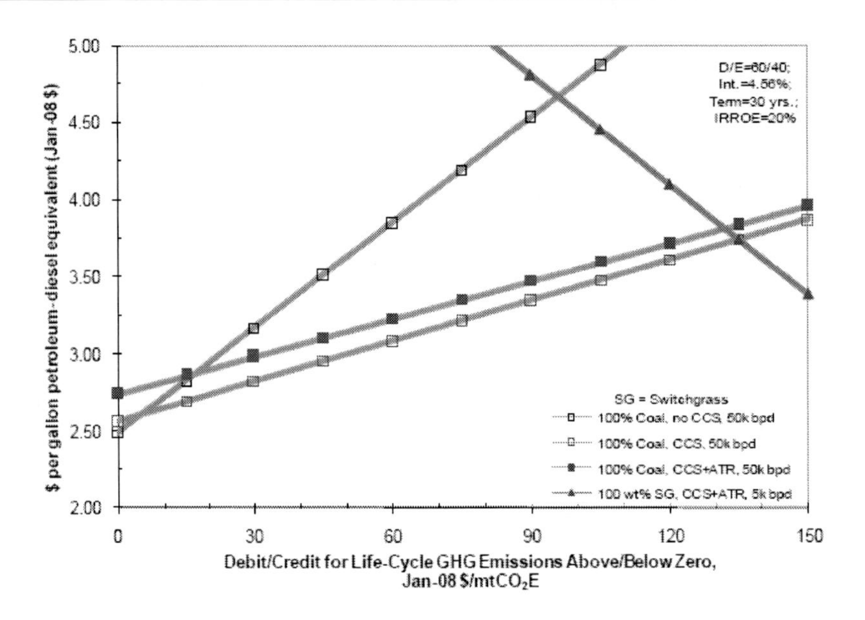

Figure 4-6. CTL Cases - Effect of GHG Emission Value on Diesel RSP.

Adding CCS to the CTL case reduces the life cycle GHG emissions associated with the fuel, resulting in a line with a less steep slope than the CTL without CCS case. Adding an ATR further reduces GHG emissions in all the CTL cases, further reducing the slope of the RSP line.

The following conclusions can be drawn from an examination of the y-intercepts on Figure 4-6:

- When the GHG emission value is zero, the CTL+CCS case is economically feasible when diesel prices are $2.56/gal or higher. This is only $0.07/gal higher than the CTL without CCS case, indicating how inexpensive it is to compress and sequester the carbon captured in the CTL plant.[30]
- When the GHG emission value is zero, the CTL+CCS+ATR case is economically feasible when diesel prices are $2.74/gal or higher. This is only $0.25/gal higher than the CTL without CCS case.

The following conclusions can be drawn from an examination of the line intersections on Figure 4-6:

- CTL without CCS is the economically preferred option when GHG emission values are below $5/mtCO2eq.
- CTL+CCS is the economically preferred option over a wide range of GHG emission values: between $5 and $135/mtCO2eq.
- With the addition of the CTL+CCS alternative, the BTL+CCS+ATR case is economically preferred only when GHG emission values exceed $135/mtCO2eq. Again, this is well above the range of GHG emission values that most regulatory forecasts contemplate.

- The incremental cost of reducing CTL GHG emissions by adding an ATR is not justified unless the GHG emission value is extraordinarily high (greater than \$285/mtCO_2eq).

4.3.3. Coal and Biomass Co-Conversion Cases

Although the previous sections showed that using biomass alone is not an economically preferred option, using biomass in conjunction with coal is much more cost-effective. Accordingly, two new cases that co-convert coal and biomass together are considered next: 15wt% CBTL+CCS and 30wt% CBTL+CCS. These cases were also considered with the addition of an ATR, but the ATR was again found to be economically justified only when the GHG emission value was extraordinarily high (\$212/mtCO_2eq or higher).

In Figure 4-7, lines are included for the 15wt% CBTL+CCS case and the 30wt% CBTL+CCS case. Cases 1, 2, and 11 (CTL without CCS, CTL with CCS, and BTL with CCS and ATR) are retained on Figure 4-7 because they continue to be the economically preferred alternatives over certain ranges of GHG emission values.

The following conclusions can be drawn from an examination of the y-intercepts on Figure 4-7:

- When the GHG emission value is zero, the 15wt% and 30wt% CBTL+CCS cases are economically feasible when diesel prices are at or above \$2.82/gal and \$3.23/gal, respectively. Note that increasing the percentage of biomass in the feed significantly increases the diesel RSP of the product when GHG emission values are relatively low. This is because biomass (switchgrass in this case) is more expensive than coal on both an energy and mass basis, and adding the ability to handle, prepare and co-feed biomass increases the capital cost. Furthermore, for the 30wt% CBTL case, limited biomass availability constrained the capacity of the plant to 30,000 bpd (as opposed to 50,000 bpd for the CTL plants and the 15wt% CBTL plant).[31]

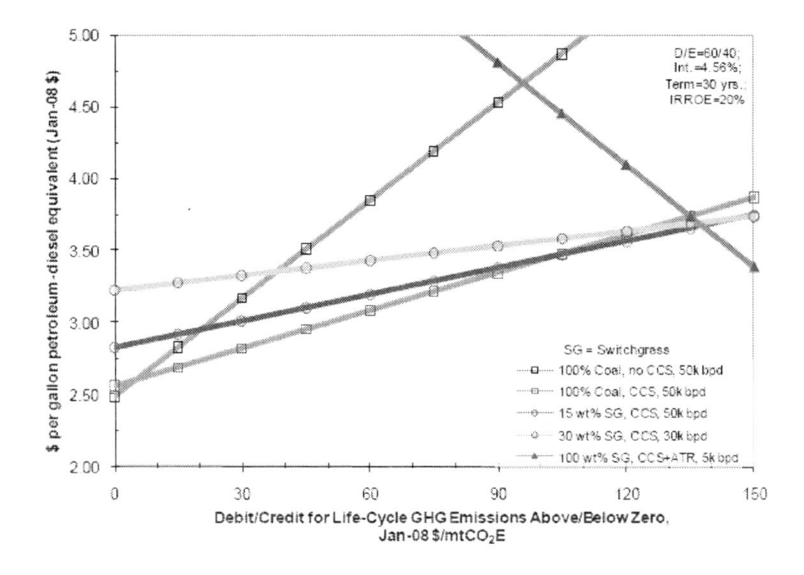

Figure 4-7. CBTL Cases - Effect of GHG Emission Value on Diesel RSP.

The following conclusions can be drawn from an examination of the line intersections on Figure 4-7:

- Increasing the biomass feed percentage from 15wt% to 30wt% is not economically justified until GHG emission values are at least \$147/mtCO_2eq.
- At GHG emission values below \$101/mtCO_2eq, the CTL or CTL+CCS cases are the economically preferred options.
- When the GHG emission value is between \$101 and \$138/mtCO_2eq, the 15wt% CBTL+CCS case is the economically preferred option.
- At GHG emission values above \$138/mtCO_2eq, the BTL+CCS+ATR case is the economically preferred option.

Although the above analysis shows that co-converting biomass with coal is not economically preferable 0 unless GHG emission values are very high, the value of CBTL becomes much more apparent if one considers the likelihood that low carbon fuel standards will be employed to force the attainment of specified GHG emission levels. This is discussed in the next section.

4.3.4. Reduced Alternatives under a Low-Carbon Fuel Standard

As shown in Table 4-1, several current and proposed regulations require certain classes of alternative transportation fuels to achieve a GHG emission profile that is a certain level below the petroleum baseline. This study refers to such a regulation as a "low carbon fuel standard" (LCFS). The requirements listed in Table 4-1 are described in more detail below and in Sections 4.4.6, 5.1 and 5.2.

Table 4-1. Current and Proposed Requirements for GHG Emission Reductions for Transportation Fuels

Source	Required Reduction below Petroleum Baseline
EISA 2007 Section 526	"below"
EISA 2007 Renewable Fuels Standard	20%
California Low Carbon Fuel Standard [a]	10% by 2020 and "more thereafter"
National Low Carbon Fuel Standard (Proposed) [b]	5% by 2015 and 10% by 2020
Senator Obama's Requirement for Near-Term CTL[c]	20%

[a] Global Warming Solutions Act of 2006, California Assembly Bill 32.

[b] First proposed in legislation introduced by Senators Obamaand Harkin in 2007 and reiterated in the "Obama-Biden New Energy Plan for America".

[c] Statement issued from Senator Obama'ssenate office on June 12, 2007.

In certain situations, the application of LCFSs would preclude the production and/or use of a non-qualifying fuel. This is in contrast to other regulatory approaches, which would allow the continued use of such fuels albeit with an economic penalty for emissions. Since the objective of an LCFS is to force the attainment of specified GHG emission levels, they intentionally preclude the option of allowing non-qualifying fuels to comply by paying a tax or purchasing allowances.

Accordingly, a LCFS could stimulate the creation of a market for fuels which meet or exceed the given standard.[32] The application of low carbon fuel standards is happening both at a federal and state level. The standards are being used to:

1. Reduce emissions from the transportation sector by reducing the carbon intensity of all transportation fuels (e.g. California's Global Warming Solutions Act of 2006 (AB-32), and legislation proposed by Senator Barack Obama in 2007),[33, 34]
2. Explicitly limit which fuels qualify for federal subsidies (EISA 2007 Renewable Fuels Standard35), and
3. Disqualify the production of other fuels based on environmental criteria.

Based on legislation currently enacted, a low-carbon fuel standard is likely to require up to a 10% or 20% reduction in limited lifecycle GHG emissions below the petroleum baseline.[36] Of the coal-only cases, only the CTL+CCS+ATR case (12% below the petroleum baseline) would qualify under a 10% LCFS. Moreover, unless fuel blending was allowed, a 20% LCFS would eliminate all of the coal-only alternatives discussed above.

President-elect Obama has stated that he would not support a CTL plant unless its life-cycle GHG emissions would be 20% less than conventional fuels (see Section 5.2) Figure 4-8 shows the economically preferred CBTL cases that could qualify under a 20% LCFS. The CTL cases have been omitted due to not meeting this standard, and a new 8wt% CBTL+CCS case has been added that precisely meets the 20% GHG emission reduction standard.

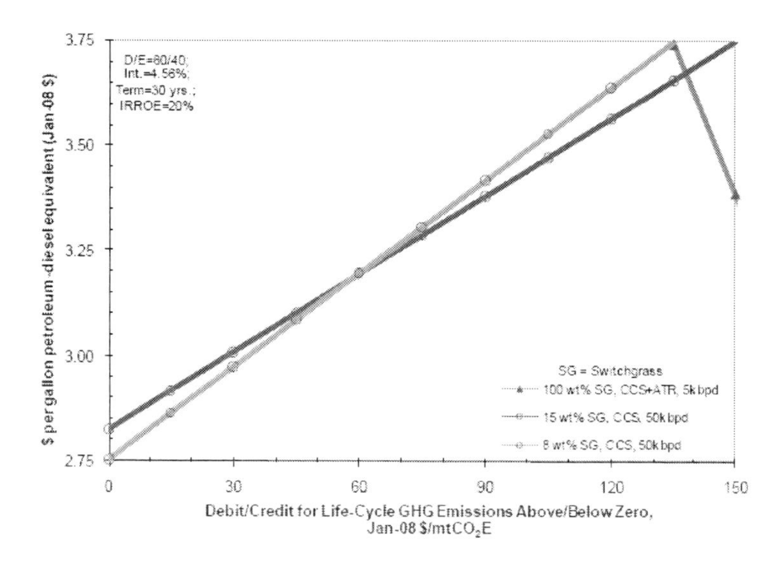

Figure 4-8. CBTL Plants that Produce Diesel Fuel with Life-Cycle GHG Emissions at Least 20% Below the Petroleum Baseline.

The following conclusions can be drawn from an examination of the y-intercepts on Figure 4-8:

- When the GHG emission value is zero, the 8wt% and 15wt% CBTL+CCS cases are economically feasible when diesel prices are at or above \$2.75/gal and \$2.82/gal, respectively.

The following conclusions can be drawn from an examination of the line intersections on Figure 4-8:

- At GHG emission values below \$58/mtCO^{2}eq, the 8wt% CBTL case is economically preferred.
- At GHG emission values above \$138/mtCO$_2$eq, the BTL+CCS+ATR case is the economically preferred option.
- The 15wt% CBTL+CCS case is the economically preferred option when the GHG emission value is between \$58 and \$138/mtCO^{2}eq.

4.3.5. CTL/CBTL/BTL Economic Feasibility Relative to Crude Oil Price

As discussed in Section 4.1, to assess whether a CTL/CBTL/BTL project will be economically feasible over its operating life, one must compare its diesel RSP to the expected future market price of ULSD (on an energy equivalent basis). Since the petroleum-derived diesel price is a function of the crude oil price, one can determine the crude oil price that would result in a petroleum-derived diesel price that is equivalent to the FT diesel RSP

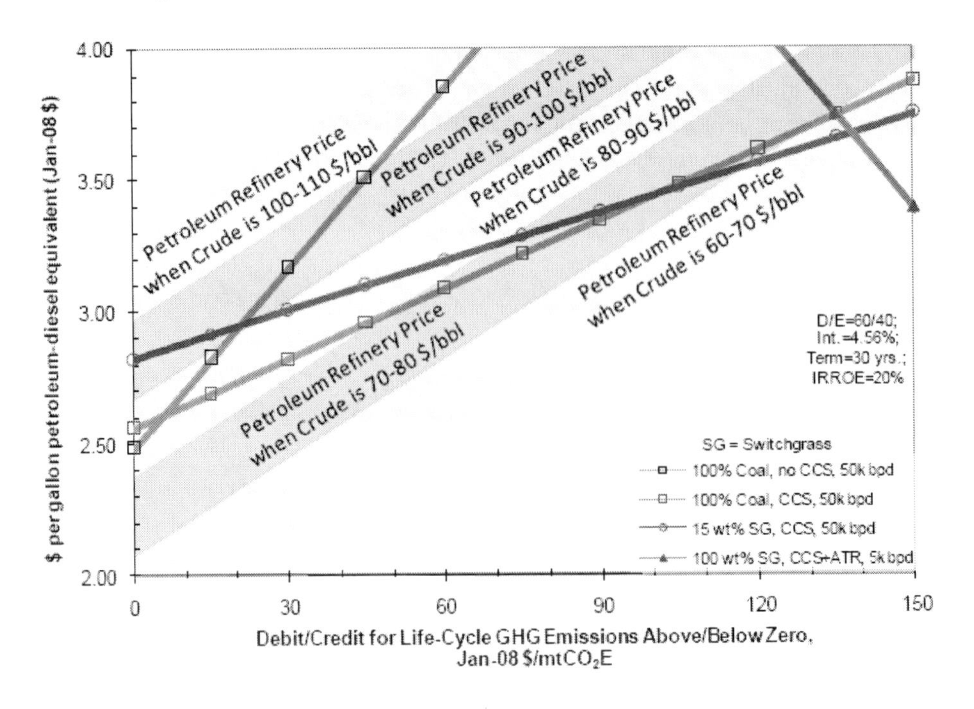

Figure 4-9. Conditions Required for Parity between the Prices of FT Diesel and Petroleum-Derived Diesel.

This is illustrated in Figure 4-9, which is a combination of previous Figures 4-1 and 4-7. It overlays FT diesel RSP curves for preferred CTL/CBTL/BTL configurations over bands that show the petroleum-derived diesel price as a function of the crude oil price. Using Figure 4-9, one can estimate the crude oil price at which each CTL/CBTL/BTL configuration would be economically feasible, i.e., the crude oil price that would result in parity between the prices of FT diesel and petroleum-derived diesel. For example, if the GHG emission value was zero (y-intercept), the CTL+CCS configuration would be economically feasible when crude oil prices are \$86/bbl or higher. Likewise, if the GHG emission value was \$60/mtCO_2eq, CTL+CCS would be feasible when crude oil prices are \$79/bbl or higher. This approach could be used to determine the minimum crude oil price required for each of the CBTL configurations to be economically feasible over the full range of GHG emission values, which is what is shown by the curves in Figure 4-10. The data shown graphically in Figure 4-10 is summarized in Table 4-2, which lists the crude oil prices required for CTL/CBTL/BTL economic feasibility when the GHG emission value is zero, along with how that value would change as the GHG emission value is increased. *Because the diesel fuel produced by any of the plants that employ CCS and/or utilize biomass has lower life-cycle GHG emissions than petroleum-derived diesel, the crude oil price required for these plants to be economically feasible actually decreases as the GHG emission value increases. In other words, the CTL/CBTL/BTL plants become more competitive with petroleum-derived diesel as the GHG emission value increases.*

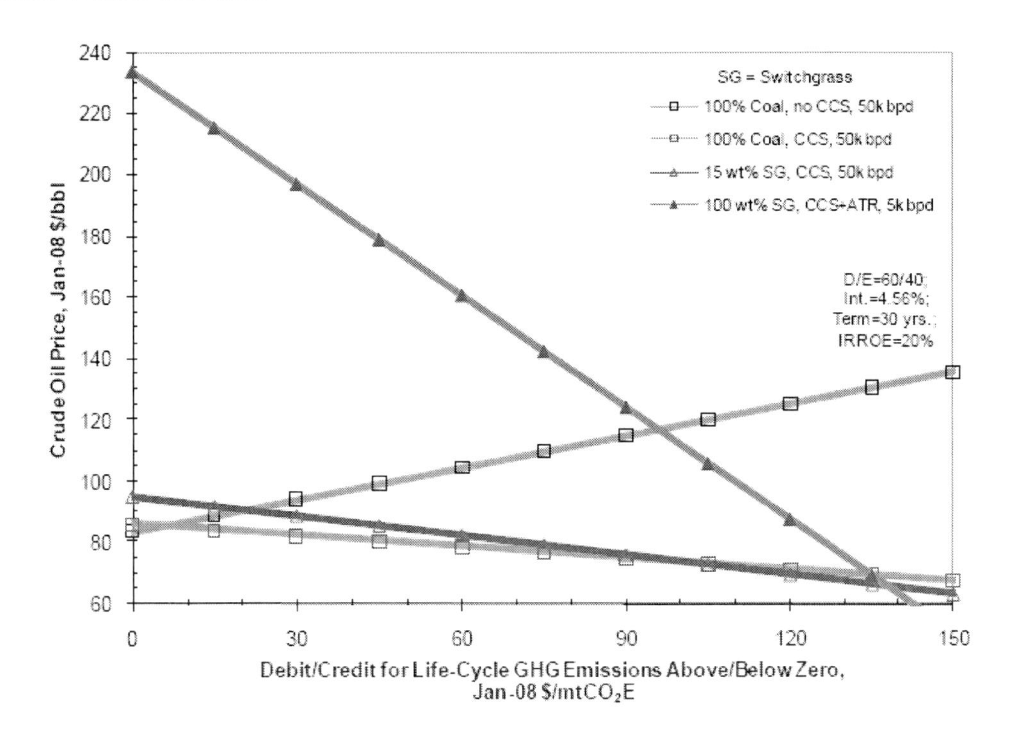

Figure 4-10. CTL/CBTL/BTL Economic Feasibility Relative to Crude Oil Price.

Table 4-2. CTL/CBTL/BTL Economic Feasibility Relative to Crude Oil Price

CBTL System	Minimum Crude Oil Price Required for Economic Feasibility (20% IRROE) when the GHG Emission Value is Zero, $/bbl	Change in Required Crude Oil Price for Every $10 Increase in GHG Emission Value, $/bbl
100% Coal, no CCS, 50k bpd	83.57	3.47
100% Coal, CCS, 50k bpd	86.08	-1.25
100% Coal, CCS+ATR, 50k bpd	92.07	-1.46
8 wt% SG, CCS, 50k bpd	92.52	-1.71
15 wt% SG, CCS, 50k bpd	94.88	-2.11
100 wt% SG, CCS, 5k bpd	225.62	-11.07

4.3.6. Summary of CTL/CBTL/BTL Economic Feasibility

For discrete ranges of GHG emission values, Table 4-3 lists which configurations would be economically preferred without a low carbon fuel standard, along with the range of crude oi l prices that would be necessary for economically feasi bi l ity, i .e. to attain an IRROE of 20%. Without a low carbon fuel standard, the CTL and CT L +CC S configurations would be economically preferred over a very wide range of G HG emission val ues ($0 to $101/mt$CO_2$eq).

Table 4-3. Preferred CBTL Alternatives for Various GHG Emission Values and Mini mum Crude Oil Price Required for Economic Feasibility

Without a Low Carbon Fuel Standard		
GHG Emission Value, Jan-08 $/mt$CO_2$eq	Preferred CBTL Alternative (lowest cost producer)	Crude Oil Price Required for Parity Between CBTL Diesel and Petroleum-Derived Diesel
0 to 5	100% Coal, no CCS, 50k bpd	$84 to $85 per bbl
5 to 101	100% Coal, CCS, 50k bpd	$85 to $74 per bbl
101 to 138	15 wt% SG, CCS, 50k bpd	$74 to $66 per bbl
138 and higher	100 wt% SG, CCS+ATR, 5k bpd	$66 and lower

President-Elect Obama has called for a National Low Carbon Fuel Standard that would requi re fuels suppliers to reduce the carbon of thei r fuel by ten percent by 2020. Table 4-4 lists which CBTL configuration would be economically preferred under a scenario in which a low-carbon fuel standard requires diesel fuel to have life-cycle GHG emissions that are 10% lower than the petroleum baseline. Under a 10% LC FS, the CTL +CCS+ATR would be economically preferred when the GHG emission value is less than \$18/mtCO_2eq. CBTL +CCS configurations fed with 8 to 15 wt% biomass would be economically preferred for a very wide range of GHG emission values, between \$18 and \$138/mtCO_2eq. GHG emission values over \$138/mtCO_2eq are not expected to be economically or political ly acceptable, since they would increase the cost of petroleum-derived diesel fuel by more than \$1.70 per gallon.

Table 4-4. Preferred CTL/CBTL/BTL Alternatives for Various GHG Emission Values and Minimum Crude Oil Price Required for Economic Feasibility: Under a 10% LCFS

Under a 10% Low Carbon Fuel Standard		
GHG Emission Value, Jan-08 \$/mtCO_2eq	Preferred CBTL Alternative (lowest cost producer)	Crude Oil Price Required for Parity Between CBTL Diesel and Petroleum-Derived Diesel
0 to 18	100% Coal, CCS+ATR, 50k bpd	\$92 to \$89 per bbl
18 to 58	8 wt% SG, CCS, 50k bpd	\$89 to \$83 per bbl
58 to 138	15 wt% SG, CCS, 50k bpd	\$83 to \$66 per bbl
138 and higher	100 wt% SG, CCS+ATR, 5k bpd	\$66 and lower

President-elect Obama has stated that he would not support a CTL plant unless its life-cycle GHG emissions would be 20% less than conventional fuels (see Section 5.2) Table 4-5 lists which configurations would be economically preferred under a scenario in which a low-carbon fuel standard requires diesel fuel to have life-cycle GHG emissions that are 20% lower than the petroleum baseline. Under a 20% LCFS, the CBTL+CCS configurations fed with 8 to 15 wt% biomass would be economically preferred over the full range of generally anticipated GHG emission values (\$0 to \$138/mtCO_2E).

Table 4-5. Preferred CTL/CBTL/BTL Alternatives for Various GHG Emission Values and Minimum Crude Oil Price Required for Economic Feasibility: Under a 20% LCFS

Under a 20% Low Carbon Fuel Standard		
GHG Emission Value, Jan-08 \$/mtCO_2eq	Preferred CBTL Alternative (lowest cost producer)	Crude Oil Price Required for Parity Between CBTL Diesel and Petroleum-Derived Diesel
0 to 58	8 wt% SG, CCS, 50k bpd	\$93 to \$83 per bbl
58 to 138	15 wt% SG, CCS, 50k bpd	\$83 to \$66 per bbl
138 and higher	100 wt% SG, CCS+ATR, 5k bpd	\$66 and lower

As discussed in Section 3.3.1, CTL/CBTL/BTL plants can produce FT diesel fuel that has life-cycle GHG emissions that are significantly reduced from the life-cycle GHG emissions of petroleum-derived diesel fuel. Greater reductions are achieved as the percentage of biomass in the feedstock is increased and/or carbon capture is enhanced by the addition of an ATR. Unfortunately, over a wide range of GHG emission values ($0 to ~$100/mtCO_2E), both of these configuration changes increase the RSP of the diesel fuel produced.

This is illustrated in Figure 4-11, which shows how CTL/CBTL/BTL configurations that achieve greater life-cycle GHG emission reductions have higher RSPs for the FT diesel fuel they produce. RSP values are plotted for GHG emission values of $0 and $60/mt$CO_2$eq. Note that although the BTL configurations [those fueled with 100% switchgrass (SG)] achieve extremely high GHG emission reductions (109 to 358% below the petroleum baseline), their diesel RSPs are so high ($5.49 to $6.96/gal) that they would only be economically feasible when crude oil prices are at least $159/bbl and higher [and even then they would only be economically preferred if the GHG emission value was also very high (greater than $138/mt$CO_2$E).]

The CTL+CCS and CBTL+CCS configurations offer what might be the most pragmatic solution: a) GHG emission reductions that are still significant (5 to 75% below the petroleum baseline), b) diesel RSPs that are only half as much ($2.56 to $3.43/gal) as the BTL cases, and c) economic feasibility when oil prices are $86 to $90/bbl.

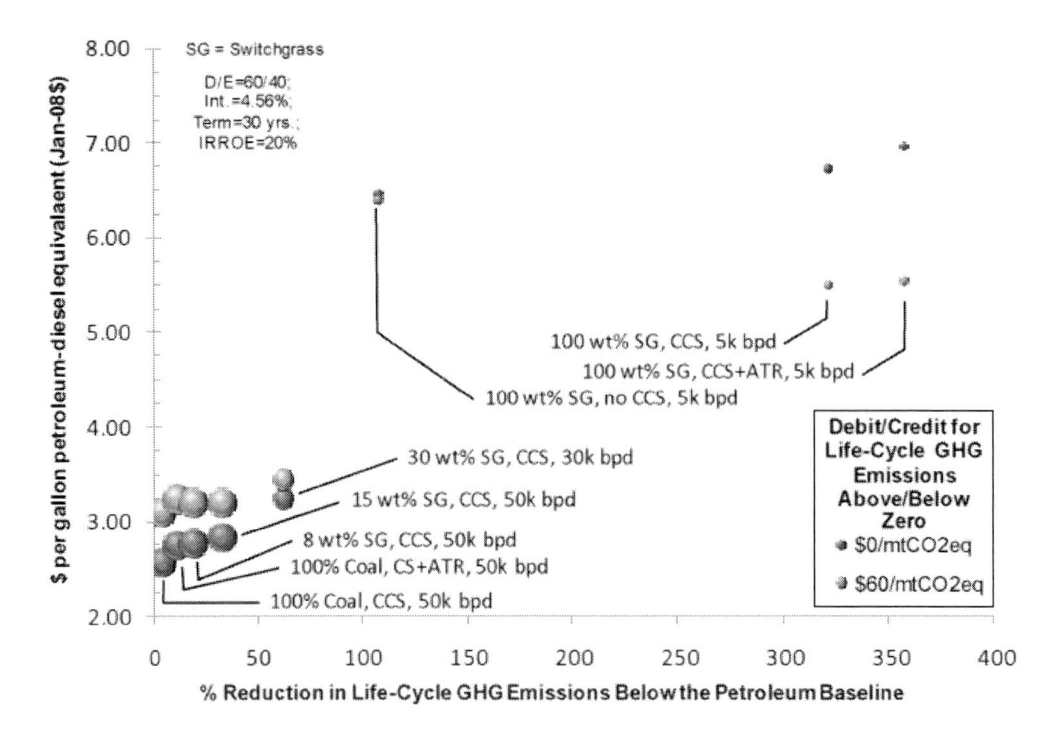

Figure 4-11. Supply Cost Curve for Low-Carbon Diesel Fuels from CTL/CBTL/BTL.

Figure 4-12 presents the same information as Figure 4-11, but at a scale that more clearly shows the data for CTL+CCS and CBTL+CCS systems.

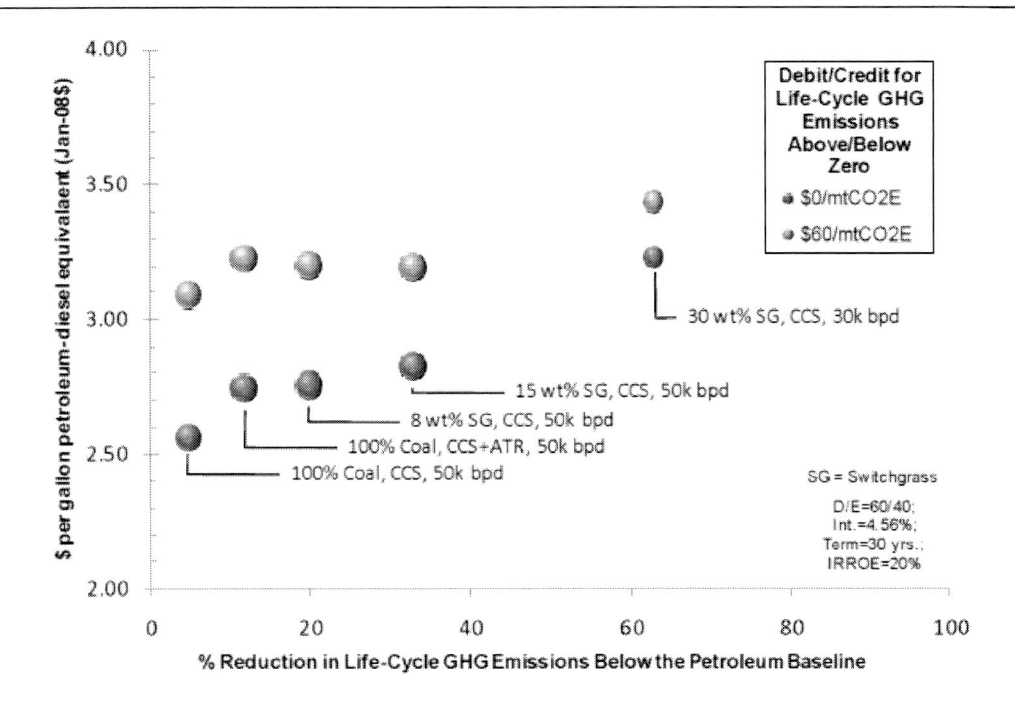

Figure 4-12. Supply Cost Curve for Low-Carbon Diesel Fuels from CTL/CBTL/BTL.

The findings of this section can be summarized as follows:

- CTL without CCS is the economically preferred alternative when GHG emission values are less than \$5/ mtCO₂eq. CTL is economically feasible over this range of GHG emission values when the market price for ULSD fuel is \$2.60/gal (i.e., when crude oil prices are \$85/bbl or higher).
- Adding CCS to CTL is economically justified when the GHG emission value exceeds \$5/mtCO₂eq and produces diesel fuel with life-cycle GHG emissions that are 5% below the petroleum baseline.
- Adding CCS to CTL is very inexpensive, increasing the diesel RSP by only \$0.07/gal. CTL+CCS is economically feasible when crude oil prices are \$86/bbl or higher when the GHG emission value is zero. At higher GHG emission values, CTL+CCS is economically feasible at even lower crude oil prices (e.g., \$79/bbl when the GHG emission value is \$60/ mtCO₂eq).
- Absent a low-carbon fuel standard, the CTL+CCS+ATR case is not an economically preferred option. However, it is the only CTL case that could meet a 10% LCFS, having life-cycle GHG emissions that are 12% below the petroleum baseline.
- Absent a low-carbon fuel standard, adding biomass to the coal feed is not an economically preferred option unless GHG emission values exceed \$101/ mtCO₂eq.
- Under a 10% LCFS, the CTL+CCS+ATR would be economically preferred when the GHG emission value is less than \$18/mtCO₂eq. CBTL+CCS configurations fed with 8 to 15 wt% biomass would be economically preferred for a very wide range of GHG emission values, between \$18 and \$138/mtCO₂eq.

- Under a 20% LCFS, configurations that co-convert coal and biomass would be economically preferred when GHG emission values range between $0 and $138/mt$CO_2$eq.
- The 15wt% CBTL+CCS plant has life-cycle GHG emissions that are 33% below the petroleum baseline and is economically feasible when crude oil prices are $95/bbl or higher when the GHG emission value is zero. At higher GHG emission values, the 15wt% CBTL+CCS plant is economically feasible at even lower crude oil prices (e.g., $82/bbl when the GHG emission value is $60/ mt$CO_2$eq).
- Adding an ATR to enhance carbon capture in CBTL plants is not economically justified unless the GHG emission value is extraordinarily high (above $212/ mt$CO_2$eq).
- With or without a 20% LCFS, BTL plants are not economically preferred unless GHG emission values are extraordinarily high (above $138/ mt$CO_2$eq).
- In the range of GHG emission values in which BTL plants are economically preferred (above $138/ mt$CO_2$eq), the addition of an ATR to enhance carbon capture is economically justified. However, GHG emission values over $138/mt$CO_2$eq are not expected to be economically or politically acceptable, since they would increase the cost of petroleum-derived diesel fuel by more than $1.70 per gallon. Consequently, there is not an expected scenario in which BTL would be economically feasible.

5. Coal and Biomass Synergies & Performance under Fuel Standards

Given a fixed amount of coal and a fixed amount of biomass, is it better to use these resources separately, in stand-alone CTL and BTL plants, or together, in CBTL plants? Which approach improves our national energy security the most? Which approach achieves the greatest reduction in GHG emissions? Which approach can produce affordable diesel fuel and still be economically feasible?

These questions are considered below under two regulatory requirements for GHG emission reductions: 1) a "Section 526" requirement (GHG emissions below the petroleum baseline), and 2) a "Renewable Fuels" requirement (GHG emissions 20% below the petroleum baseline).

The findings show that under either regulatory requirement, the synergistic use of coal and biomass in CBTL plants is economically preferable to, and achieves greater GHG emission reductions than, using these resources separately in stand-alone CTL and BTL plants. Under the Renewable Fuels emission requirement, the synergistic CBTL approach has a much greater impact on energy security, enabling the economic production of twenty times more diesel fuel from secure, domestic energy resources. When used together in CBTL plants, coal benefits by the environmental synergy afforded by co-gasifying biomass, and biomass benefits by the economic synergy afforded by co-gasifying coal. Without these synergies, neither of these domestic fuels could be utilized in a manner that was both economically feasible and environmentally acceptable under the Renewable Fuels requirement.

5.1. Section 526 Requirement

Section 526 of the Energy Independence and Security Act of 2007 (EISA 2007) prohibits federal agencies from procuring a fuel unless its life-cycle GHG emissions are equal to or less than those for conventional petroleum sources (the “petroleum baseline”).[37] Table 5-1 describes two scenarios: one in which coal and biomass are used in separate CBTL plants (CTL and BTL plants) to produce FT diesel fuel, and a second scenario where coal and biomass are co-gasified for the synergistic production of FT diesel fuel.

In each scenario, the economically preferred plant configuration within the respective CTL, BTL and CBTL “families” was chosen based on the ability to meet the Section 526 criteria (reduced GHG emissions compared to the petroleum baseline) and a GHG emission value between $30 and $60 per metric ton CO_2-equivalent.[38] The following plant configurations were thereby selected:

Case 2 (CTL+CCS): 5% below the petroleum baseline
Case 4 (8 wt% CBTL+CCS): 20% below the petroleum baseline
Case 10 (BTL+CCS): 322% below the petroleum baseline

Table 5-1. CBTL Pathways with Life-Cycle GHG Emissions Lower than Petroleum Derived Diesel Fuel (Section 526 Requirement)

APPROACH	PATHWAYS	Biomass Consumed, million short-tons/year (dry)	Coal Consumed, million short-tons/year (as-received)	# of plants	Total Synthetic Diesel Produced, million bbl/year	Reduction of Life-Cycle GHG Emissions Below the Petroleum Baseline		Investment NPV* (Billions Jan-2008$) when Crude Oil Price is $100/bbl and GHG Emission Reduction Credit is Below Value	
						%	million mtCO₂eq per year	$30 per mtCO₂eq	$60 per mtCO₂eq
Separate Use of Coal and Biomass	CTL w/CCS	0.0	118.5	17	191	5	4.5	10.5	12.8
	BTL w/CCS	9.2	0.0	8	9	322	13.0	-2.9	-1.6
	TOTAL / Weighted Avg	9.2	118.5		200	19	17.5	7.6	11.1
Synergistic Use of Coal and Biomass	8 wt% CBTL w/CCS	8.8	118.0	18	203	20	19.1	7.9	11.2

*Net Present Value at a discount rate of 20%, assuming the following project finance structure: 60% debt at an annual nominal interest rate of 4.56% and a term of 30 years.

Under the Section 526 Requirement, the synergistic use of coal and biomass together in CBTL plants results in slightly higher GHG emission reductions.

5.1.1. Separate Use of Coal and Biomass

If 118 million short-tons/year of coal and 9 million short-tons/year of biomass were separately consumed in 17 CTL+CCS plants and 8 BTL+CCS plants, a total of 200 million bbl/year of diesel fuel would be produced and GHG emissions would be reduced below the petroleum baseline by 17.5 million $mtCO_2eq$/year.

Both the CTL+CCS diesel fuel and the BTL+CCS diesel fuel would meet the Section 526 requirement by having life-cycle GHG emissions less than the petroleum baseline (5% below and 322% below, respectively). If the decision was made to blend the products of both plants together, the resulting fuel mixture would have life-cycle GHG emissions that are 19% below the petroleum baseline, just short of the Renewable Fuels requirement.)

The total NPV for the two projects would range from $7.6 to $11.2 billion when the credit for GHG emission reductions ranges from $30 to $60 per metric ton CO_2-equivalent (even though the NPV for BTL plants would be negative).

5.1.2. Synergistic Use of Coal and Biomass

Alternatively, if the same amounts of coal and biomass (118 and 9 million short-tons/year) were consumed together in 8 wt% CBTL plants, about the plant outputs and revenue would be about the same as if these resources were used separately: a similar amount diesel fuel would be produced, about the same GHG emission reductions would be achieved and the NPV would remain roughly the same.

5.1.3. Summary

Under a "Section 526" regulation that requires FT fuels to have life-cycle GHG emission levels at or below the petroleum baseline, the synergistic use of coal and biomass resources in CBTL plants would result in somewhat lower GHG emissions compared to the separate use of coal and biomass in CTL and BTL plants. Furthermore, the CBTL option provides a way to use biomass and still achieve a strongly positive NPV, unlike BTL projects, which are not economically feasible (negative NPV).

Under a "Section 526" regulation, the use of biomass for diesel production in BTL plants is not economic in this analysis. However, the environmental benefits of biomass can still be captured if the economic synergy of co-gasifying biomass with coal in a CBTL plant is employed.

5.2. Renewable Fuels Requirement

Fuels produced from new biorefineries must have life-cycle GHG emissions twenty percent lower than the life-cycle emissions from petroleum-derived diesel in order to qualify as a renewable fuel under Title II, Subtitle A of EISA 2007.[39] One might think of this as a 20% low carbon fuel standard.

A similar requirement has been proposed for coal-derived fuels. A statement issued from his senate office on June 12, 2007 stated that "Senator Obama supports research into all technologies to help solve our climate change and energy dependence problems, including shifting our energy use to renewable fuels and investing in technology that could make coal a clean-burning source of energy. However, unless and until this technology is perfected, Senator Obama will not support the development of any CTL fuels unless they emit at least 20% less life-cycle carbon than conventional fuels [22]."

By combining coal with 8 wt% biomass in a CBTL+CCS plant, a 20% low carbon fuel standard could be achieved. However, no CTL configuration can meet this emission requirement -- even if an ATR were added to enhance carbon capture, a CTL+CCS+ATR system would only reduce life-cycle GHG emissions to twelve percent below the petroleum baseline.

5.2.1. Separate Use of Coal and Biomass

As shown in Table 5-2, if 9 million short-tons/year of biomass were consumed in 8 BTL+CCS plants, 9 million barrels per year (bpy) of diesel fuel would be produced and GHG emissions would be reduced below the petroleum baseline by 13 million mtCO_2eq/year. The reduced output – 9 million bpy compared to the 200 million bpy shown in Table 5-1 is due to the fact that coal cannot be used to produce diesel fuel under this "separate use" approach, as no CTL plant can meet the 20% reduction in GHG emissions criteria. This also impacts the total GHG reduction, as the CTL+CCS plants in Table 5-1 contributed to the total emissions reductions.[40]

The total NPV is always negative in this "separate use" approach, and would range from -$2.9 to -$1.6 billion when the credit for GHG emission reductions ranges from $30 to $60 per metric ton CO_2-equivalent.

Table 5-2. CBTL Pathways with Life-Cycle GHG Emissions at Least Twenty Percent Lower than Petroleum Derived Diesel Fuel (Renewable Fuels Requirement)

APPROACH	PATHWAYS	Biomass Consumed, million short-tons/year (dry)	Coal Consumed, million short-tons/year (as-received)	# of plants	Total Synthetic Diesel Produced, million bbl/year	Reduction of Life-Cycle GHG Emissions Below the Petroleum Baseline		Investment NPV* (Billions Jan-2008$) when Crude Oil Price is $100/bbl and GHG Emission Reduction Credit is Below Value	
						%	million mtCO_2eq per year	$30 per mt$CO_2$eq	$60 per mt$CO_2$eq
Separate Use of Coal and Biomass	CTL w/CCS	– CANNOT MEET REQUIREMENT –							
	BTL w/CCS	9.2	0.0	8	9	322	13.0	-2.9	-1.6
Synergistic Use of Coal and Biomass	8 wt% CBTL w/CCS	8.8	118.0	18	203	20	19.1	7.9	11.2

*Net Present Value at a discount rate of 20%, assuming the following project finance structure: 60% debt at an annual nominal interest rate of 4.56% and a term of 30 years.

Under a 20% low carbon fuel standard, the synergistic use of coal and biomass together in CBTL plants achieves far greater reductions in GHG emissions while producing twenty times more diesel fuel and garnering positive (rather than negative) economic returns.

5.2.2. Synergistic Use of Coal and Biomass

If the same amount of biomass were consumed together with 118 million short-tons/year of coal in an 8 wt% CBTL+CCS plant, the level of diesel fuel production would be 20 times higher and nearly 50% more GHG emission reductions would be achieved. Moreover, the total NPV would be strongly positive, indicating that the economics of CBTL plants would remain very attractive in this scenario.

5.2.3. Summary

Under a standard that requires FT fuels to have life-cycle GHG emission levels that are 20% below the petroleum baseline, the synergistic use of coal and biomass resources in CBTL plants would enable biomass to be used in an economically feasible manner and would result in much lower GHG emissions compared to an approach that attempted to convert coal and biomass in separate plants. *Moreover, the CBTL approach has a much greater impact on energy security, enabling the economic production of 20 times more diesel fuel from secure, domestic energy resources.*

If one assumes that BTL plants with negative NPVs would not be built, the advantages of using coal and biomass together in CBTL plants are even greater. If alternative fuels are required to have GHG emissions 20% below the petroleum baseline, CTL plants would be prohibited by the emission requirement and BTL plants would be economically infeasible, leaving CBTL as the remaining option for producing diesel fuel from these secure, domestic resources.

Under a 20% low carbon fuel standard, coal cannot be used without the environmental synergy afforded by co-gasifying biomass, and biomass cannot be used without the economic synergy afforded by co-gasifying with coal.

Under a 20% low carbon fuel standard, coal benefits by the environmental synergy afforded by co-gasifying biomass, and biomass benefits by the economic synergy afforded by co-gasifying coal. Without these synergies, neither of these domestic fuels could be utilized in a manner that was both environmentally acceptable and economically feasible.

6. Potential Economic Benefits of a Large CBTL Industry

This chapter delineates the potential economic benefits that would accrue to the United States should it develop a large scale CTL/CBTL industry to both substantially improve energy security while reducing CO_2 emissions. Analysis has shown that CTL/CBTL can be competitive at oil prices far exceeded by recent levels, and at levels projected in the recently issued IEA World *Energy Outlook 2008* [23]. This publication forecast 2030 oil prices to be double the forecast in *WEO 2006,* (in 2030, \$120/bbl versus \$55/bbl, with 10 million barrels of day (mmb/d) less consumed, globally (106 mmb/d versus 116 mmb/d) [23, 24]. Under this oil price projection the NPV, in 2008, of a 3mmb/d – 5 mmb/d CBTL industry would be in the range of \$200 billion - \$700 billion.

6.1. Oil Price Projections

The reason for the new perspective on oil prices is the dearth of "cheap oil," and, despite the financial collapse of 2008, the realization that rekindled economic growth will soon lead to oil prices that meet or exceed the commonly accepted 2008 cost of incremental oil, \$70-\$80/bbl. The IEA projects that continued consumption growth in Asia and the Middle East will more than offset any consumption decline in the West, and projects 2010 oil prices of \$100/bbl. (See Figure 6-1, which shows projections adjusted to 2008 price levels). Of course, in late 2008 oil prices collapsed to the \$40- \$50/bbl range, far off their July 2008 peak exceeding \$140/bbl. However, for prices to languish in this range for more than a few years would be indicative of deep Western economic malaise and anemic Asian economic growth.

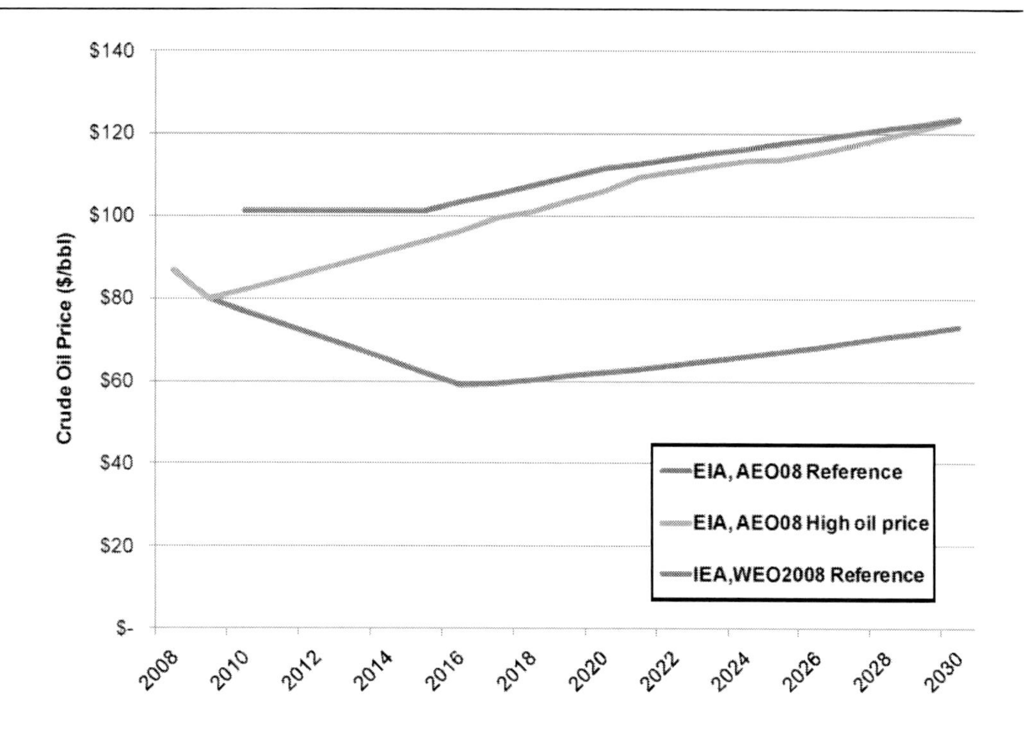

Figure 6-1. Current Oil Price Projections, $ per barrel, year 2008 prices.

Table 6-1. RSP, Crude Oil Equivalence

CO_2 charges	CTL w/o CCS	CTL+CCS	CTL+CCS+ ATR	8wt% CBTL+CCS	15wt% CBTL+CCS	BTL w/o CCS	BTL+CCS+ ATR
CO_2 = $0/t	$84.50	$86.58	$92.52	$93.09	$95.44	$218.43	$234.94
CO_2= $45/t	$100.09	$80.98	$85.97	$85.41	$85.95	$198.29	$180.11
CO_2 = $90/t	$115.69	$75.37	$79.43	$77.74	$76.43	$178.14	$125.27

The fundamental difference between the energy crisis of the 1973-1985 and that of 2003-2008 is that the latter was caused primarily by a large increase in demand paired with resource difficulty, worsened by financial speculation in 2008, rather than by supply disruptions caused by the voluntary withholding or by revolutionary upheaval [25]. Thus, once the business cycle turns, the forces of growth will increase oil demand and with it prices. In the meantime, a period of low oil prices due to economic stagnation provides a window of opportunity to develop new energy alternatives that could be deployed for future sustained periods of high prices, a more prudent option than waiting for the next crisis.[41]

The micro-level systems engineering analysis of this report demonstrates that the RSPs of coal/biomass diesel to be well below the level projected by IEA, in terms of crude oil equivalence (Table 6-1, first row). The analysis in Chapter 4 further demonstrates that CTL is preferred to coal-biomass blends at moderate carbon prices. That said, under a potential low carbon fuel standard, CBTL options are preferred to biomass-only alternatives under any reasonable carbon price scenario.

6.2. Economic Profits

It is important to note that the RSP indicates the level at which all expected returns to labor and capital have been met; that is, investors have achieved profits commensurate with their expectation (in this study, a 20% IRROE). In economics jargon, the project pays a "normal" rate of return or an accounting profit. An economic profit is defined as a return to capital above the normal rate. Therefore, at a price equal to the RSP, economic profit is zero. When economic profits are positive, an economic "rent" has been earned, in this case because the world oil price is above the RSP. At or above such levels the economic case for the competitive domestic alternative becomes compelling, based on otherwise forgone economic profit, obtainable increases in net surplus, amelioration of the trade deficit, and job creation.[42] At prices below the RSP, the economic profit is negative and the CTL/CBTL industry would be in need of some type of subsidy for the country to achieve the energy security and trade balance benefits associated with this industry. However, a long period of low oil prices, indicative of economic stagnation, would see as well a decline in the capital costs from the January 2008 levels embedded in the current RSPs, bringing down the RSPs associated with the different CTL/CBTL plants and improving economic competitiveness.

The economic benefits of CTL/CBTL may even be greater than those that would be obtained under the IEA price projection. There is a risk that the IEA projection is low, based on the difficulty with which remaining oil resources are extracted or the reluctance of oil-producing nations to adequately invest. In this case, the lack of supply growth would result in even higher prices, further limiting consumption growth. This is borne out with the following: a theoretical projection of oil prices of $150/bbl by 2030 would increase prices by less than 25% (relative to the "IEA Reference" case).[43] Note that between the IEA *WEO2006* and *WEO2008* projections, a 100% increase in price elicits less than a 10% decline in consumption. With this level of demand responsiveness, the further increase in price of 25% (i.e. $150/bbl by 2030) would only limit world oil consumption to about 103 mmb/d. Therefore, the high oil price sensitivity of an admittedly speculative projection to $150/bbl highlights the insurance value of CTL/CBTL under very high oil prices, since large demand for CTL/CTBL diesel fuel would still exist.

In Figure 6-2, the area shaded under the IEA Reference price projection indicates the economic profits attained by CTL/CBTL. The area above that curve and below the NETL high oil price projection is additional economic profit under higher oil prices. The economic profits retained under these assumptions are assumed to accrue to domestic owners, representing economic rent not transferred overseas due to the displacement of imports. However, these levels should be reduced by the "propensity to import", that is, the portion of consumption spent on imports. This ratio is 0.16, that is, for every $1 of economic profit, 84 cents is retained, to be shared by capital owners and governments.

To calculate those areas under the curve, a ramp up of the CTL/CBTL industry must be assumed. Compared to levels of production projected by Energy Information Administration (EIA) or International Energy Agency (IEA), a large scale ramp-up would require significant incentives, at least for the first few plants, until oil price uncertainty is reduced and technical risk assuaged. While multiple scenarios could be envisioned, Figure 6-3 illustrates those used for purposes of these calculations.[44] These ramp-ups are for comparison purposes only; they are not based on any particular readiness or impact study. The comparisons in the chart are to EIA's AEO2008 high oil price case and to the IEA reference case [26].

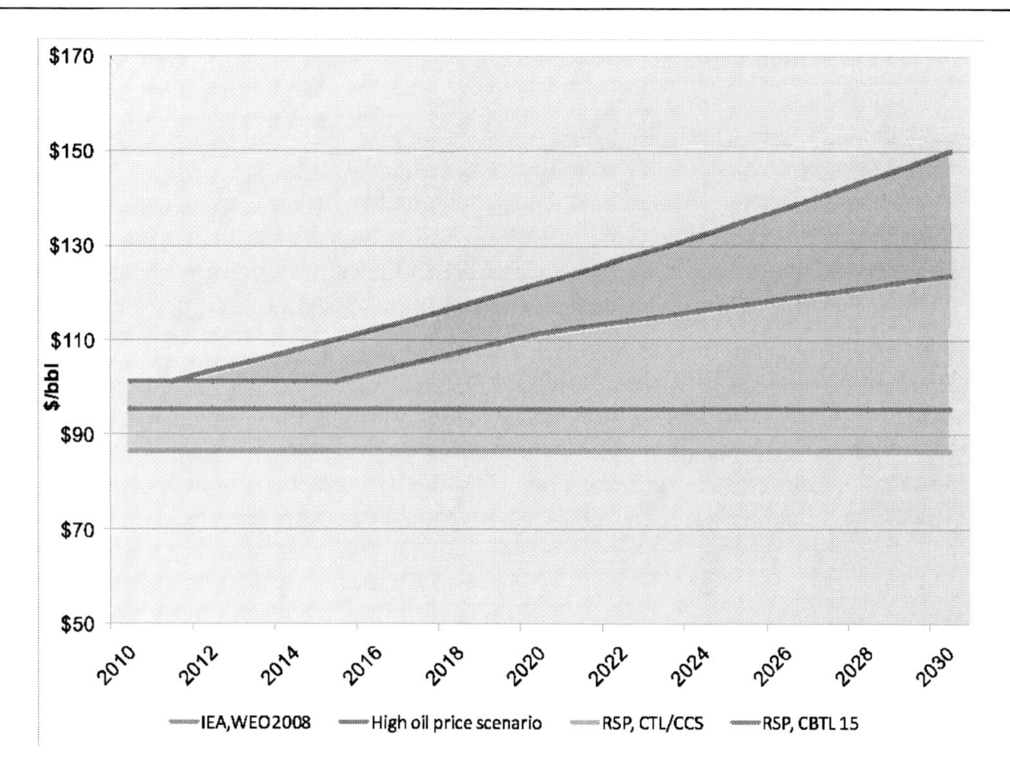

Figure 6-2. Economic Profits under Oil Price Projections.

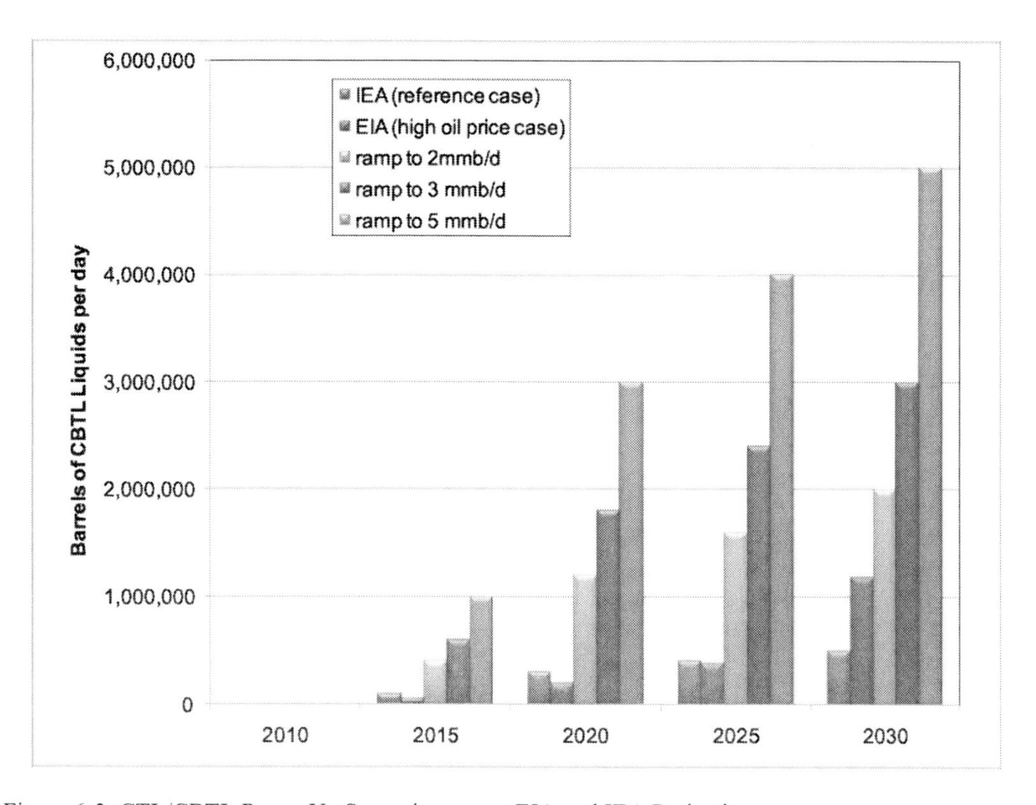

Figure 6-3. CTL/CBTL Ramp-Up Scenarios versus EIA and IEA Projections.

6.3. Effect on World Oil Prices

Standing up an industry of this scale will likely have an effect on the world price of oil. The RAND Corporation, in a new report, develops a plausible world oil market model: RAND concludes that, for each million bpd of alternative fuels, in this case CTL/CBTL, world oil prices will decline from 0.6% to 1.6% [27]. The higher percentage figure is adopted here, for, as RAND points out, the larger price effect will occur in a scenario of high world oil prices, which the IEA reference case certainly is. Table 6-2 shows these percentage effects under the IEA projection and under the high oil price sensitivity.

Further, RAND argues that the decline of world oil prices *per se* will generate a net surplus to the United States of $6/b to $30/b, with the latter figure obtaining when the world oil price is high [27]. The word net is important, because the figure accounts for lost revenue by domestic oil producers. This effect is not technology-specific, but would benefit the US if, for any reason, oil imports were reduced by the magnitudes contemplated here. The economic benefits shown here thus use the upper bound, yielding an annual benefit exceeding $30 billion in 2030, for a 3 mmb/d industry, and over $90 billion for a 5 mmb/d industry. In a low world oil price world, with prices below the RSP, the required subsidy would still be offset by a figure equivalent to 20% of the benefits calculated here.

Table 6-2. Effect of Ramp-up on World Oil Prices

World Oil Prices	2020	2030
IEA reference	$111.56	$123.73
2 mmb/d ramp	$109.95	$120.76
3 mmb/d ramp	$108.88	$118.78
5 mmb/d ramp	$106.74	$114.82
High Oil Price Scenario	$122.08	$150.00
2 mmb/d ramp	$120.32	$146.40
3 mmb/d ramp	$119.15	$144.00
5 mmb/d ramp	$116.80	$139.20

6.4. Trade Deficit Issues

The effect on the trade deficit is an issue that excites much debate. Some assert nearly a one to one relationship; that is, for every dollar not spent on imports, Gross Domestic Product (GDP) increases by that dollar, whereas others disclaim any positive effect, which is akin to asserting the trade balance is always net zero.[45] Projecting trade deficit effect is quite uncertain, since the nation's terms of trade and appetite for imports are factors difficult to establish for future periods. Nonetheless since net exports are a component of GDP, we observe that for every dollar not sent overseas, foreign purchases of US exports will be reduced by some fraction. Between 1969 and 2007, exports averaged 85% of imports; between 1999-2008, 70%; between 2002-2007 for the Middle East, 60%. A range of benefit determined by the excess of imports could thus be quite wide. Here, the 70% offset figure is used; that is, for every $1 reduction in imports, GDP grows by 30 cents.

6.5. Job Creation

Finally, the topic of job creation should be addressed. In a 2006 report for NETL, the authors employ an economic input-output model to determine that, approximately, 150,000 jobs would be created for every million bpd of production. Whether these jobs are net additions to employment levels or represents shifts between sectors continues to be debated.[46]

In the near-term, however, the United States is of course not in a full-employment economy. Over the longer term, since CTL and CBTL would necessarily involve a large construction effort and employ many highly technical workers, a large ramp up would probably create a premium for labor that might reorient the income distribution in favor of labor, similar to the experience in the Canadian oil sands. This might bump the RSP upwards by increasing the cost of CTL/CBTL plant builds. This assertion and other labor market issues are beyond the scope of this report, except to simply note that the growth of a high value-added manufacturing industry in the United States would be a positive development.

6.6. Summary

The potential benefits to the United States of CTL/CBTL under a regime of high oil prices could be vast. For a 3 mmb/d industry, economic benefits could exceed $100 billion dollars on an annual basis by 2030. *On a NPV basis for the period 2010-2030, the value of the industry in 2008 dollars approaches $400 billion.* Table 6-3 shows both the annual benefit in 2030 and the NPV of the stream of benefits from 2010 to 2030. Figures of this magnitude- and indeed any component thereof- indicate the massive public benefit potentially available to the United States. With the possible exception of oil shale, it is difficult to see any other domestic supply alternative or supplement to conventional oil at such scale. Moreover, although CBTL is not conventional oil, it is compatible with both the current liquids fuel infrastructure and also with leading demand side alternatives, such as plug-in, diesel-electric hybrid automobiles. In a carbon constrained world, the added cost of CCS to the CTL plant is merely $5/ton CO_2eq, or equivalently an incremental $2/bbl, clearly identifying CTL+CCS as a leading candidate for CCS demonstrations under an aggressive climate policy. Further, if paired with prospective EOR, CTL+CCS would replace imports with two thrusts and reduce lifecycle GHG emissions, dramatically so in CBTL applications.

Table 6-3. Summary: Potential economic benefits of CTL Options, by 2030. (Million $2008

	3 mmb/d, 2030	5 mmb/d, 2030	NPV,* 3 mmb/d	NPV, 5 mmb/d
Retained Economic Profit,	*Annual*	*Annual*	*2008-2030*	*2008-2030*
Reference Case				
CTL+CCS	$29,698	$43,411	$88,331	$132,448
15wt% CBTL+CCS	$21,527	$29,792	$55,601	$77,899
Additional economic profit, **high oil price projection**	$23,257	$37,470	$52,849	$85,925
Reduced world oil prices	$32,850	$91,250	$84,980	$236,056
Trade deficit amelioration				
IEA Reference	$39,228	$47,556	$146,164	$238,318
High Oil Price case, incremental	$23,973	$29,062	$18,925	$30,770
Sums				
CTL+CCS, Reference	$101,776	$182,217	$319,474	$606,822
CTL+CCS, HOP	$149,005	$248,749	$391,249	$723,516
15wt% CBTL+CCS, Reference	$93,605	$168,598	$286,744	$552,272
15wt% CBTL+CCS, HOP	$140,834	$235,130	$358,519	$668,966
Job creation (cumulative)			300,000-400,000	500,000

* NPV = Net Present Value, calculated at a 7% discount rate. This rate exceeds the commonly accepted value of 6% for longterm returns on common stock in the 20th century; a lower rate of , say, 4% implies public good benefits not displayed. A positive value indicates a benefit to the United States.

7. RESEARCH & DEVELOPMENT NEEDS

Significant research, development and demonstrations have been conducted on the gasification of pure coals and of pure biomass. Coal and biomass mixtures have also been researched, but there is a minimal amount of data available on gasification of different coal ranks with different biomass mixtures at different ratios. Additionally, FT synthesis for the production of liquid transportation fuels has been demonstrated on large scales historically for reasons including market instability, national vulnerability and limited national resources. However, several uncertainties and questions need to be addressed, and these can be grouped into three key research areas: (1) bench and engineering-level development and demonstration, (2) systems analyses, and (3) site specific design studies.[47] In response to Congressional direction in the FY2008 budget appropriations language, the United States DOE's Office of Fossil Energy and National Energy Technology Laboratory are aggressively pursuing R&D in areas (1) and (2), with the aim of reducing the economic and technical challenges associated with large-scale deployment of CBTL projects and promoting the widespread acceptance of this method of fuels production. Area (3) is identified as an activity that could help guide future R&D and systems analysis efforts because there is a lack of detailed engineering design for CTL and CBTL plants.

7.1. Bench and Engineering Development & Demonstration

The first component of a strategy consists of a research and development activity to identify and address the technical challenges associated with co-feeding different types of coal and biomass at varying feed mixture percentages.

The bench and engineering-level research is needed to:

a) demonstrate methods to successfully introduce a coal-biomass feedstock into the high pressure - high temperature gasifier regime;
b) perform a complete bench-scale characterization of the effluent gas products resulting from the gasification of several coal/biomass combinations, and [48]
c) determine the amounts of trace contaminant species expected from coal/biomass mixtures that the FT and WGS processes can tolerate to ensure effective and economical operation.

Subsequent research and development in all of these areas will be necessary to provide final proof-of-concept for control of contaminants, successful feeding systems, optimal, integrated and sustained operation in pilot facilities.

Specific areas for R&D include, but are not limited to:

1. What is the influence of co-feeding various ranks of coal and types of biomass at different ratios on the kinetics of gasification related reactions?

Specific information related to the influence of co-feeding different coal and biomass ratios, as well as different species, must be generated. While systems analysis studies have predicted that co-feeding biomass at 8-15wt% provides unique carbon emissions and economic advantages, the presence and the nature of the biomass is sure to have an effect on the gasification process and reactions, most likely due to differences in heating value; oxygen, carbon and hydrogen content; and the presence of metals associated with the biomass to catalyze the gasification reactions.

2. What is the influence of co-feeding various ranks of coal and types of biomass at different ratios on the solid, liquid and gaseous products resulting from gasification?

The presence of biomass will affect the gasification conditions and may result in different types of species being contained in the solid, liquid, and gaseous products resulting from gasification. Temperature differences caused by the presence of biomass may alter the amounts of residues produced relative to those from pure coal gasification. Moreover, biomass will introduce a unique contaminant stream, which may alter the composition of the solids, liquids, and gases produced.

3. What are the optimum gasifier operating conditions for the conversion of coal and biomass to transportation fuels?

Historically, pure coal and pure biomass gasification research has been conducted at different temperatures; biomass typically being conducted at lower temperatures as compared to coal. In a co-gasification configuration, the different heating value, moisture content and elemental species associated with biomass may require non-traditional operating conditions in order to maximize gasifier and process efficiency.

4. How does the variability of biomass species and feed ratios influence the reliability of sustained operation?

It is possible that the presence of biomass will alter the reliability of long term operation, primarily due to the variability in biomass content and species, and the influence of co-derived syngas on the gasifer and down-stream processes.

5. What is the optimum pretreatment of biomass required for feeding into a high-pressure gasifier? *The gasification characteristics of biomass are different than those of coal, and it follows that it may be desirable to pre-treat the biomass in order to more closely match the characteristics of coal. Pretreatment processes including drying, sieving, pelletizing, torrefaction or grinding, may be beneficial in maintaining consistent operation of the gasifier under a variety of co-feeding scenarios as well as increasing the energy density of the biomass, harvest and distribution area, and storage properties.*

6. What is the optimum feed arrangement for coal and biomass in the co-gasification process? *Although large scale gasification has been practiced on pure coal and pure biomass, it is unclear what feed configuration would result in the optimum operation of a CBTL process: (1) pre-mixing the coal and biomass and feeding at one point, or (2) two distinct trains (one for biomass and one for coal) feed at two different locations within the gasifier, etc.*

7. What is the optimum configuration of the gasifier(s)?

Due to the potential difference in reaction rates of coal and biomass as well as the variability of biomass supply and species, various gasifier configurations can be explored and optimized; a single gasifier for both coal and biomass or separate gasifiers for coal and biomass arranged in series or parallel.

8. What is the influence of co-gasification products on the materials associated with the gasifier? *The difference in elemental content of biomass as compared to coal may impact the service life of gasifier materials as well as down-stream processes. For example, will the high alkali content of the biomass impact the slag-refractory interactions within the gasifier?*

9. What is the influence of co-gasification products on syngas processing equipment?

The primary and trace constituents resulting from gasification of biomass and coal may affect syngas processing equipment and chemistry (e.g. FT and WGS catalysts). For example, biomass will introduce alkali species and it is not known at what levels these species will begin to affect processes and processing equipment. It will also be important to determine any synergistic effects, both positive and negative, that multiple contaminant species have on the overall process.

10. Do current commercial cleaning processes remove the trace syngas species associated with co-gasification?

It is unclear which contaminants will be present and at what levels they will be introduced by various species of biomass. To some degree, this will be specific to the particular species of biomass and the feed rate. However, certain syngas cleaning processes will be more likely to also remove trace contaminants than others. This will affect removal capacity, rate of absorption of target contaminants versus competitive uptake of co-contaminants, and also any regeneration processes.

7.2. Systems Analysis

The second research area would incorporate systems analyses to evaluate technologies and R&D progress and provide guidance for the research activities. Five key areas of systems analyses for CBTL plants are:

- *Techno-economic analyses* to assess and define the state of technology and RD&D progress and needs; *Risk analyses* to identify, characterize, and evaluate the technical, economic and environmental hurdles associated with deploying CBTL technologies and the strategies required to overcome them;
- *Resource and infrastructure analyses* to identify, define and offer strategies to address the critical "readiness" issues – availability of engineering and design firms, skilled labor, railroad and other transportation capacity, mining capability, biomass availability and its impacts on plant capacity and operations, materials and equipment availability and other potential critical needs;
- *Benefits analyses* to determine the advantages of introducing CBTL technologies as one element in the suite of alternative fuel options; and
- *Market analyses* to evaluate commercial application of technologies.

This study is the first major CBTL systems study performed by NETL and will be followed on with other work which evaluates the other key issues listed above.

7.3. Site-Specific Design Studies

The third research area could support site-specific designs of first-of-a-kind CBTL plants, including front-end engineering designs, to prove the feasibility of achieving near-zero GHG emissions. The purpose of this technical area is to collect detailed engineering design information, which is currently lacking, to guide future R&D and systems analysis efforts and to provide industry with experience and confidence in the technology.

Advanced coal gasification and FT conversion technologies have been developed to reduce product cost, but have not been demonstrated in an integrated system at sufficient scale to confirm the potential economics and efficiencies of production. Recently, baseline systems designs and associated construction cost analyses have been updated for CTL plants and initial analyses have recently been published and others are forthcoming for CBTL plants.

Although these analyses represent best engineering judgment, an overwhelming amount of risk will continue to exist until plants that integrate the technologies to produce liquid fuels from coal and coal/biomass mixtures are designed, built, and operated in the United States. Several CTL projects have been announced in the United States and worldwide which, if developed, could form a database of information on the technology and reduce the technical and financial risk. The only CBTL project announced to date is by Baard Energy LLC. They plan to design and construct a 50,000 BPD CBTL plant in Wellsville (Columbiana County), Ohio and have recently announced that they are starting the Front End Engineering Design (FEED) phase for the plant [31]. It is anticipated that DOE will closely follow the progress being made on this activity and other planned first-of-a-kind commercial CTL and CBTL plants.

To facilitate the deployment of CBTL plants, several site-specific CBTL FEED studies could be performed on representative coals of different rank (bituminous, sub-bituminous or lignite), and different types of biomass (switchgrass, corn stover, poplars, forest residues, and mixed prairie grasses). Various feed concentrations of coal and biomass having diverse characteristics would be evaluated in these studies to provide a basis for commercial validation, decisions to proceed, and financing. The completed designs at different site locations would provide industry with the knowledge and experience to consider further commercial deployment.

8. CONCLUSIONS

This study evaluates the use of the United States' abundant domestic resources to address the often competing priorities of energy security, climate change mitigation, and economic sustainability. A comprehensive assessment was performed to evaluate the economic feasibility and climate change impact of coal and/or biomass conversion to diesel fuel using a process known as indirect liquefaction, and more specifically the FischerTropsch catalytic synthesis process combined with carbon capture and sequestration.

The key findings of this study were that:

- By coupling two existing technologies, indirect liquefaction and carbon sequestration, coal and biomass can be economically converted into ultra-low sulfur diesel fuel which has significantly less life cycle GHG emissions (5 to 75% less) than petroleum-derived diesel fuel. For example, conversion of 8% (by weight, 5% by energy) biomass with coal can result in a 20% reduction in life cycle GHG emissions when CCS is employed.
- Despite a fairly costly financial structure (20% IRR on equity), CTL/CBTL diesel fuel is competitive with petroleum-derived diesel when oil prices are as low as \$86/bbl; should carbon constraints become the norm, these fuels would become more

competitive as carbon prices increase (because they have lower life-cycle GHG emissions).

- When coal and biomass are used together in CBTL plants, the coal feedstock benefits by the environmental synergy afforded by co-gasifying biomass, and the biomass feedstock benefits by the economic synergy afforded by co-gasifying coal. Without these synergies, neither of these domestic fuels could be utilized in a manner that was both economically feasible and environmentally acceptable if such fuels are required to achieve life-cycle GHG emissions that are substantially lower (>12%) than the petroleum baseline.
- CTL/CBTL plants offer the least-cost option to demonstrate the geologic sequestration of carbon dioxide, which is critically needed to enable our nation to continue using its valuable coal resources in a carbon-constrained world. Capturing and compressing over 90% of the carbon dioxide emissions from a coal-toliquids plant is inexpensive (adding only seven cents per gallon to the product cost) and the resulting carbon dioxide can be pumped into our nation's older oil wells to recover large volumes of leftover, difficult-to-extract oil, while simultaneously locking the carbon dioxide underground.

Furthermore, a national commitment to promote the use of CTL and CBTL would have a tremendously positive impact on the economy, creating skilled jobs and reducing the amount of money sent overseas for oil imports, valued at $326 billion in 2007 and between $400 and $500 billion in 2008. The production of domestic diesel would also improve the economic competitiveness of domestic industries by easing supply constraints associated with diesel fuel, thereby reducing overhead costs associated with high fuel costs. Should oil prices resume their upward trend, the economic and security benefits of CBTL to the nation could be enormous.

Based on these findings, CTL and CBTL represent a well balanced and pragmatic solution to the United State's energy strategy dilemma of achieving energy security and addressing climate change in an economically sustainable manner. Biomass to Liquids (BTL) - only becomes economically competitive when the GHG emission value exceeds \$138/mtCO_2E and does not result in greater reductions in net GHG emissions than if the same amount of biomass were used in a CBTL plant

Of the plant configuration options evaluated, the CTL with CCS configuration and CBTL with CCS configuration (featuring 8 to 15 wt% biomass) will likely represent the optimal solutions, considering trade-offs between GHG mitigation and economic advantage. These plant configurations produce a diesel fuel which has 5% to 33% less life cycle GHG emissions than petroleum-derived diesel and can be economically competitive at diesel prices of $2.56/gallon of petroleum-diesel equivalent.[49] Inherent in these prices is a relatively stringent return on equity of 20%. Since these prices also assume transportation fuel emissions are not taxed (a GHG emissions value of zero), higher GHG emissions values consequently lead to economical competitiveness at even lower petroleum-diesel equivalent prices. Given that future oil prices are predicted to reach $130/bbl by 2030[50], these CTL/CBTL fuels would be very profitable in the future [23].

Technology Readiness and the Development Opportunities

The CTL/CBTL pathway has the technical credibility in that it is a commercially proven technology with large-scale operating experience: existing CTL plants produce 150,000 bpd of liquid fuels products, and the co-gasification of up to 30% (by weight) biomass with coal has been demonstrated in a large scale, commercially available gasifier. Nevertheless, significant R&D opportunities exist to improve the economic competitiveness and climate change benefits of this technology. These range from co-gasification research in probabilistic methods of determining the kinetics for different biomass types and mixtures, optimization of biomass processing and feeding technologies, and investigation into catalytic synthesis and hydrocarbon product processing, using Fischer-Tropsch catalysis or other methods, in order to improve product slate flexibility and process efficiency.

CTL/CBTL is therefore a near-term (3 to 5 years, based on permitting and plant construction) technology pathway, but also a pathway which can result in a new, high-tech industry. A national commitment to CTL/CBTL would result in high-tech jobs associated with not only the construction of plants, but also the plant operation and further development and optimization of CTL/CBTL component technologies, including overall plant research.

The Path Forward

The CCS technology intended to be applied to the CTL/CBTL process is critical to the future use of coal not only in this process but also in the electric utility industry. Because the use of coal is important in maintaining the competitiveness of the U.S. energy mix, this technology combination offers a unique advantage to the nation. CTL/CBTL with CCS represents the lowest-cost option to demonstrate carbon capture and storage at significant scale while adding only $2/bbl to the required selling price of the product.

CTL/CBTL w/CCS has numerous benefits, including but not limited to: (1) production of fuels with significant reductions in GHG emissions, (2) creation of a large scale domestic industry with numerous new skilled job opportunities created, (3) opportunities for R&D – leading to an important new high-tech industry, and (4) competitiveness in a key transportation fuel, using an abundant domestic energy resource, with an improving economic advantage as GHG emission values increase.

While the decline in world oil price, at the end of 2008, may temporarily preclude the economic competitiveness of CTL/CBTL fuels, long-term oil price projections show that these fuels are likely to be competitive by the time a plant can be built and extremely profitable in the long therm. As described in Chapter 6, a 3 million barrel per day industry could have domestic economic benefits which exceed $100 billion dollars on an annual basis by 2030. On a net present value basis for the period 2010-2030, the value of the industry in 2008 dollars *approaches $400 billion.* This economic potential could be enhanced under a number of scenarios, including reduced engineering, procurement, and construction (EPC) costs and commodity and equipment costs, associated with the current global recession; technological improvements to the CTL/CBTL process, or the commercial sale of the CO_2 as a byproduct, for Enhanced Oil Recovery (EOR) or other purposes.

CTL/CBTL with CCS is therefore a technology pathway that can uniquely and simultaneously provide a solution to the divergent energy objectives of our nation.

REFERENCES

[1] Energy Information Administration, "U.S. Energy-Related Carbon Dioxide Emissions by Fossil Fuel," May 2008, http://www.eia.doe.gov/oiaf/1605/flash/flash.html.
[2] Energy Information Administration, "2009 Annual Energy Outlook," Table A18, December 2008, http://www.eia.doe.gov/oiaf/aeo/pdf/appa.pdf.
[3] Energy Information Administration, "Annual Energy Review for 2007," Petroleum Statistics, June 23, 2008, http://www.eia.doe.gov/emeu/aer/petro.html.
[4] van Dongen, Ad and Marco Kanaar, "Co-gasification at the Buggenum IGCC power plant," DGMK Fachbereichstagung, April 2006,
[5] http://www.dgmk.de/kohle/abstracts velen7/Dongen Kanaar.pdf.
[6] Alleman, Teresa L. and Robert L. McCormick, "Fischer-Tropsch Diesel Fuels – Properties and Exhaust Emissions: a Literature Review," National Renewable Energy Laboratory, March 2003,
[7] http://www.sae.org/technical/papers/2003-01-0763.
[8] Ortiz, David S., Henry H. Willis, Ash Pathak, Preethi Sama, James T. Bartis, "Characterization of Biomass Feedstocks," RAND Corporation, May 2007.
[9] Reed, Michael, Jay Ratafia-Brown, Michael Rutkowski, John Haslbeck and Timothy J. Skone, "Assessment of Technologies for Co-Converting Coal and Biomass to a Clean Syngas – Task Report 2 (RDS)," National Energy Technology Laboratory, March 7, 2008.
[10] McDaniel, John, "Biomass Co-gasification at Polk Power Station, Final Technical Report," May 2002, http://www.osti.gov/bridge/servlets/purl/823831-bTDn1G/native/823831.PDF.
[11] van Dongen, Ad and Marco Kanaar, "Co-gasification at the Buggenum IGCC Power Plant," DGMK Fachbereichstagung, April 2006,
[12] http://www.dgmk.de/kohle/abstracts velen7/Dongen Kanaar.pdf.
[13] Statoil, "Carbon dioxide storage prized," Statoil, December 18, 2000, http://www.statoil.com/en/TechnologyInnovation/NewEnergy/Co$_2$Management/Pages/SleipnerVest.aspx
[14] CO2 Capture and Storage R,D & D Projects Database, http://www.co$_2$captureandstorage.info/project specific.php?project id=70.
[15] IEA GHG Weyburn-Midale CO2 Monitoring and Storage Project, December 2008, http://www.ptrc.ca/siteimages/WMP brochure revised Dec19.2008.pdf.
[16] Kreutz, Thomas G., Eric D. Larson, Guangjian Liu, and Robert H. Williams, "Fischer-Tropsch Fuels from Coal and Biomass," Princeton Environmental Institute, August 21, 2008 (revised
[17] October 7, 2008), http://www.princeton.edu/pei/energy/
[18] U.S. Environmental Protection Agency, "Identifying Opportunities for Methane Recovery at U.S. Coal Mines: Profiles of Selected Gassy Underground Coal Mines 1999-2003," September 2005, http://www.epa.gov/cmop/docs/profiles 2003 final.pdf.
[19] Milbrandt, A., "A Geographic Perspective on the Current Biomass Resource Availability in the United States," National Renewable Energy Laboratory, December 2005,
[20] http://www.nrel.gov/docs/fy06osti/39181.pdf.

[21] Skone, Timothy J. and Kristin Gerdes, "Development of Baseline Data and Analysis of Life Cycle Greenhouse Gas Emissions of Petroleum-Based Fuels," National Energy Technology Laboratory, November 26, 2008, http://www.netl.doe.gov/energy-analyses/pubs/NETL%20LCA%20Petroleum-Based%20Fuels%20Nov%202008.pdf.

[22] Tilman, D., J. Hill, C. Lehman, "Carbon-Negative Biofuels from Low-Input High-Diversity Grassland Biomass," *Science*, Vol. 314, 2006, pp. 1598-1600, http://www.sciencemag.org/cgi/content/abstract/314/5805/1598.

[23] Williamson, David M. and Alan J. Sachs, "Renewable Fuel Standard Update," Beveridge & Diamond PC, February 4, 2008, http://www.bdlaw.com/news-news-270.html.

[24] California Climate Change Portal, "Assembly Bill 32 – The Global Warming Solution Act of 2006," http://www.climatechange.ca.gov/ab32/index.html.

[25] "Obama Introduces National Low-Carbon Fuel Standard," Renewable Energy World.com, May 9, 2007, http://www.renewableenergyworld.com/rea/news/story?id=48427.

[26] "Barrack Obama and Joe Biden: New Energy for America," August 2008, http://www.barackobama.com/pdf/factsheet energy speech 080308.pdf.

[27] Wallsten, Peter, "Obama yields to greener side," *Los Angeles Times*, June 13, 2007, http://articles.latimes.com/2007/jun/13/nation/na-energypol13.

[28] International Energy Agency, "World Energy Outlook 2008," 2008, http://www.worldenergyoutlook.org/2008.asp.

[29] International Energy Agency, "World Energy Outlook 2006," 2006, http://www.worldenergyoutlook.org/2006.asp.

[30] Stevens, Paul, *The Coming Oil Supply Crunch*, The Royal Institute of International Affairs, Chatham House, London, UK, 2008.

[31] Energy Information Administration, "Annual Energy Outlook 2008," June 2008, http://www.eia.doe.gov/oiaf/archive/aeo08/index.html.

[32] Bartis, James T., Frank Camm, and David S. Ortiz, "Producing Liquid Fuels from Coal – Prospects and Policy Issues," RAND Corporation, 2008,

[33] http://www.rand.org/pubs/monographs/2008/RAND MG754.pdf.

[34] The Southern States Energy Board, "American Energy Security – Building a Bridge to energy Independence and to a Sustainable Energy Future," Norcross, Georgia, July 2006,

[35] http://www.americanenergysecurity.org/AES%20Report.pdf.

[36] Drummond, Charles, Robert L. Hirsch, Roger H. Bezdek and Robert M. Wendling, "Economic Impacts of U.S. Liquid Fuel Mitigation Options," National Energy Technology Laboratory, July

[37] 8, 2006, http://www.netl.doe.gov/energyanalyses/pubs/Economic%20 Impacts%20of%20U.S.%20Liquid%20Fuel%20Mitigation%20Opti ons.pdf.

[38] Batis, James T., Tom LaTourette, Lloyd Dixon, D.J. Peterson, and Gary Cecchine, "Oil Shale Development in the United States – Prospects and Policy Issues," RAND Corporation, Arlington, VA, 2005, http://www.rand.org/pubs/monographs/2005/RAND MG414.pdf.

[39] Baard Energy, "AMEC Paragon keeps U.S. $5 billion alternative fuel project moving," Houston, TX, October 2, 2007, http://www.baardenergy.com/press/2007-10-02%20Baard%20AMEC%20Release.pdf.

A. ECONOMIC ANALYSIS METHODOLOGY

The economic analyses presented in this report were performed using the NETL Power Systems Financial Model (a discounted cash flow analysis tool) using the following assumptions (except where noted) regarding: project finance structure, construction schedule, plant capacity factor, feedstock acquisition costs, product prices, escalation, taxes and depreciation.

A.1. Selection of a Financing Structure for Cbtl Projects

Described in Table A-1 are four scenarios for financing CBTL projects. Scenarios 1 and 1A reflect today's situation in which CBTL plants face a high regulatory risk because of uncertainty about a future carbon regulation and a high technical risk because no commercial-scale plants have been demonstrated yet in the United States. Scenario 1 is assumed to have no government incentives while extensive government incentives are assumed to be available under Scenario 1A.

Table A-1. Project Financing Scenarios

Scenario	1	1A	2	2A (basis of this study)
Timeframe	TODAY		MID-TERM	
Regulatory Risk	HIGH (no carbon regulation)		LOW (carbon regulation in place)	
Technical Risk	HIGH (no commercial-scale CBTL plants in U.S.)		MEDIUM (2 or 3 commercial-scale CBTL plants	
Market Risk	HIGH		HIGH	
Government Incentives	NONE	EXTENSIVE (e.g., cost-sharing, price guarantees)	NONE	MODERATE (e.g., loan guarantees)
Debt/Equity Ratio	0/100	?	50/50	60/40
Debt Interest Rate, %	n/a	?	8.15 (LIBOR+6)	4.56 (CMT+0.22)
Debt Term, years	n/a	?	15	30
Required IRROE (over 30 years), %	20	?	20	20

All rates are expressed in nominal terms. The average annual inflation rate is assumed to 2%.
Internal Rate of Return on Equity (IRROE) over 30-year economic life of plant.
On 11/13/2008, the 12-month London Interbank Offered Rate (LIBOR) was 2.15% and the 30-year constant-maturity Treasury (CMT) rate was 4.34%.

Scenarios 2 and 2A reflect a hypothetical mid-term future in which regulatory risk has been eliminated by the passage of a carbon regulation and technical risk has been partially mitigated by the demonstration of two or three commercial-scale CBTL plants. Scenario 2 is assumed to have no government incentives while moderate government incentives are assumed to be available under Scenario 2A.

All scenarios face a high degree of market risk because of the volatile nature of energy prices, especially the world oil price.

All the finance structures listed in Table A-1 are based on "project" financing, i.e., non-recourse financing in which debt and equity returns are paid back from the cash flow generated by the project. In project financing, non-recourse debt is secured by a pledge of collateral – typically the real assets of the project – and liability is limited to that project collateral. This is in contrast to "corporate financing" in which the corporation would hold a general liability for the amount of the loan.

The financial structures listed for Scenarios 1 and 2 are taken directly from a 2008 NETL study that recommends financial structures for analysts to use when performing economic assessments of fossil-based energy projects. [1] The recommendations were developed by Nexant, an energy sector consulting firm, based on interviews with project developers/ owners, financial organizations and law firms.

Since there is currently no carbon regulation, it is highly unlikely that any CTL or CBTL plant could obtain bank (debt) financing, meaning that 100% of project capital must come from equity sources absent any government incentives (Scenario 1). Of course, with extensive government intervention (Scenario 1A), the finance structure could change in any number of ways.

After a carbon regulation is enacted, it is assumed that debt financing will become available for CBTL projects even without government incentives (Scenario 2). The 2008 NETL study estimated that up to 50% of the project capital could be debt-financed at an interest rate that is six points higher than the twelve-month London Interbank Offered Rate (LIBOR), with the balance of capital finance through equity that requires an internal rate of return of 20%.

If government incentives were available, a more favorable finance structure could be obtained. The finance structure assumed for Scenario 2A assumes that a government loan guarantee is secured under the program authorized by Title XVII of the Energy Policy Act of 2005. Although the terms of loans issued under that program will vary, this analysis assumes that 100% of the debt portion of the project financing is guaranteed, with a loan term of 30 years (the maximum allowed) and an interest rate that is 22 basis points above the 30-year constant-maturities Treasury (CMT) rate.

The debt service coverage ratio (DSCR) is the ratio of the operating profit to the cost of debt service (principal plus interest). According to the 2008 NETL study, a DSCR above 1.75 or 2.0 will likely be required to secure debt financing for fossil-based energy projects. Accordingly, this analysis limited the percentage of capital financed with debt to 60% such that the DSCR remained above 2.0 when averaged over the first five years of plant operation. [1]

The effect of project finance structure on the diesel required selling price for various CBTL configurations is shown in Figure A-1. As shown, financing options can have a significant effect on RSP, and therefore plant viability. Of particular note is that government incentives, such as loan guarantees, can make the use of CCS of co-gasification of biomass the economically preferred option even when CO_2 emissions do not have a value: CBTL options with CCS and up to 15wt% biomass are can all produce diesel fuel for under \$3/gallon of petroleum-diesel equivalent when loan guarantees are available for these options, which is less than the \$3.10/gallon of diesel for CTL without CCS (assuming loan guarantees are not available for this option).

A.2. Construction Schedule

Construction is assumed to be completed over a period of three years (2009, 2010 and 2011), during which the annual capital cost distribution is spread evenly in real terms. Loan draws are made annually to cover construction costs, with interest accruing only on the amount drawn.

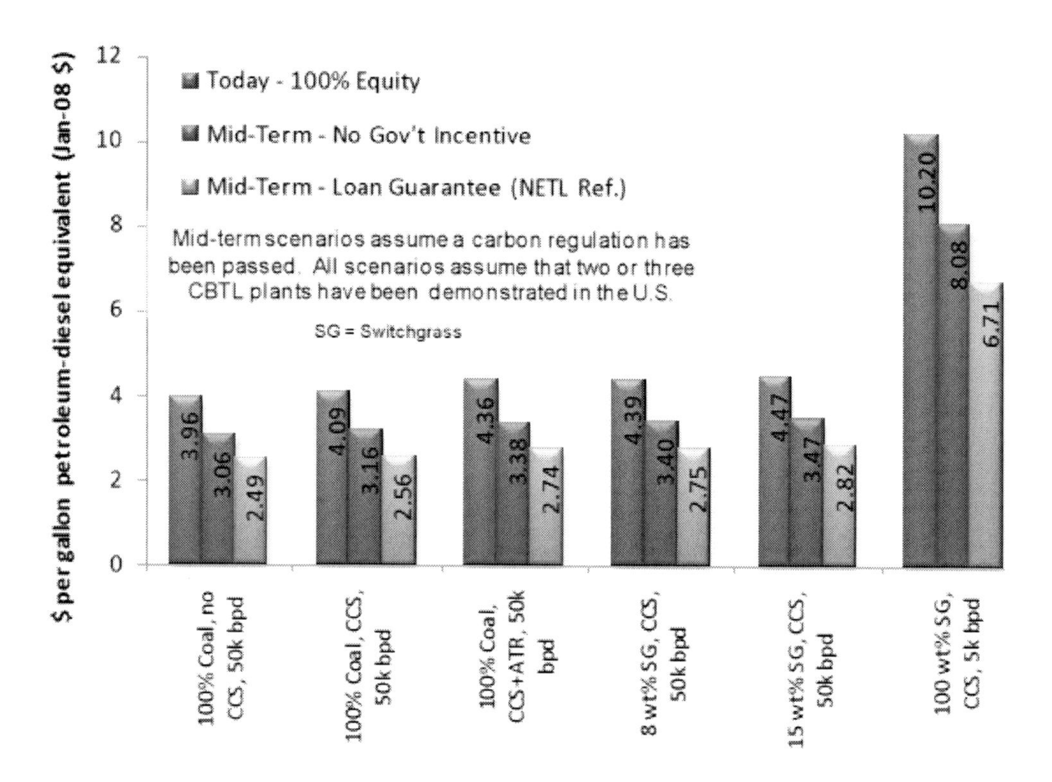

Figure A-1. Effect of Project Finance Structure on Diesel RSP.

A.3. Plant Capacity Factor

The capacity factors assumed for this study are listed in Table A -3.

Table A-3. Capacity Factor Assumptions

Plant Type	Year 1	Year 2	Years 3+
CT L and CBTL	69%	79%	90%
BTL	65%	75%	85%

Since the CTL/C BTL plants considered in this study include a spare gasifier, and since they are assumed to be has third- or fourth-of-a-kind, a design capacity factor of 90% is assumed to be achieved in the third year of operation.

This value was adjusted from a recent E PRI projection [2] of the availability that will be achieved by the next generation of Integrated Gasification Combined Cycle (IGCC) plants when no spare gasifier is included. (Since CBTL plants are assumed to be dispatched whenever they are available, the capacity factor is assumed to equal the availability.)

BTL plants are assumed to achieve a design capacity factor of 85% in the third year of operation. Although the BTL plants also feature a spare gasifier, a lower capacity factor was assumed because the supply of biomass feedstock is expected to be less reliable than the supply of coal. Seasonal weather variations can affect the magnitude, timing and quality of available biomass feedstocks. It is assumed that CBTL plants will be able to cope with such variations by using coal to compensate for shortfalls in biomass availability, but BTL plants will not have this option.

For all the CBTL plants, it was assumed that the capacity factor was below the design value in years one and two. The assumed capacity factor ramp-up rate was adjusted from the availability ramp-up rate projected by EPRI [3] for the next generation of IGCC plants. The first- and second-year capacity factors for CTL/CBTL plants are assumed to be 69% and 79%. The first- and second-year capacity factors for BTL plants are assumed to be 65% and 75%.

A.4 Feedstock Acquisition Costs

A.4.1. Coal Cost

The coal type assumed for this study is Illinois #6, a high-sulfur bituminous coal.

Reference case data [4] for the DOE Energy Information Administration's (EIA) "Annual Energy Outlook 2008" projects that the minemouth (underground) cost of high sulfur coal in the East Interior Supply Region (which includes Illinois) will be $1.34/MMBtu (year 2006 dollars) in 2012, the assumed first year of operation for plants in this study. The average transportation cost of this coal to plants within the East North Central Demand Region (which also includes Illinois) is $0.44/MMBtu. Thus the total delivered cost of the coal is projected to be $1.78/MMBtu in 2012 (year 2006 dollars).

Applying a nominal annual escalation rate of 2%, the delivered cost equates to $1.85/MMBtu in Jan-2008 dollars. For Illinois #6 coal (11,666 Btu/lb), this translates to $43.16 per short ton (Jan-2008 dollars).

A.4.2. Switchgrass Cost

NETL sponsored RAND to characterize a variety of potential biomass feedstocks, including the development of models to estimate their costs [3]. Switchgrass costs are plotted in Figure A-2 as a function of the required feed rate. The costs are for delivery to the plant gate and include costs associated with crop establishment, cultivation, harvesting and storage, and transportation.

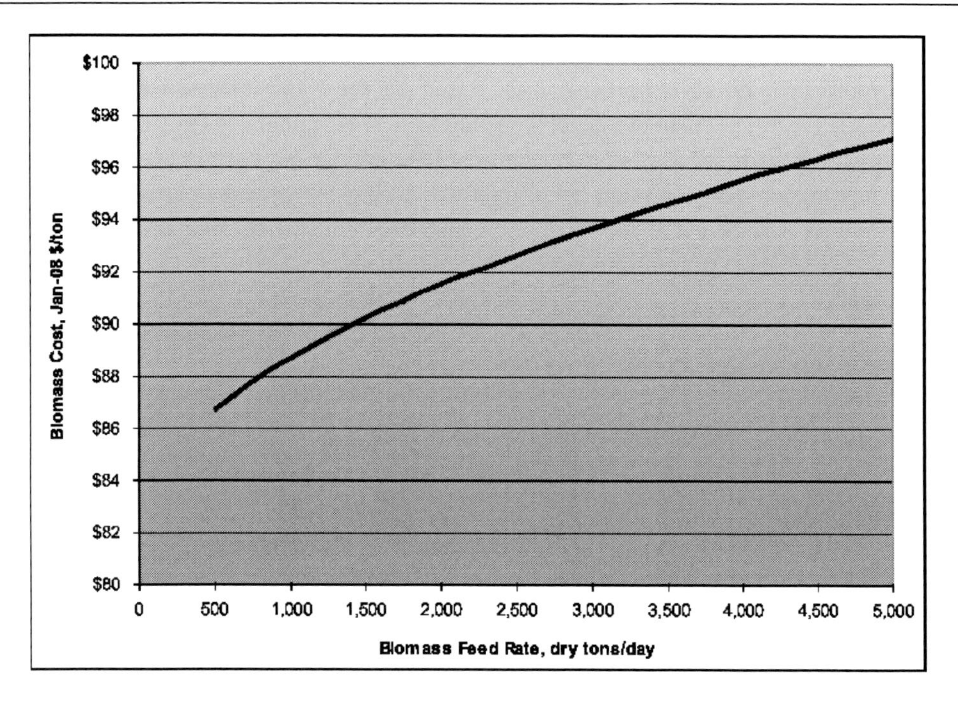

Figure A-1. Delivered Switchgrass Cost as a Function of Feed Rate.

The reported switchgrass costs assume that only marginal lands are employed for switchgrass cultivation. It is assumed that the switchgrass is field dried and stored in the field until it is needed. Final switchgrass processing is assumed to occur at the CBTL/BTL facility and is not included in the delivered cost estimates. Convention dictates that the delivered switchgrass cost be expressed in $/dry- ton (January 2008$) although the biomass is still "wet" when delivered to the plant.[51]

A.5. Product Prices

A.5.1. Diesel-Naphtha Price Ratio

Since CBTL plants produce naphtha as a byproduct, it is necessary to assume a price ratio for these products to perform the economic analysis. For this study the diesel-naphtha price ratio was assumed to be 1.3, based on communications with individuals involved in the development of a potential commercial-scale CTL project in the United States [5].

A.5.2. Diesel-Crude Oil Price Ratio

As discussed in Section 4.2, this study assumes that the ratio of the ultra-low-sulfur diesel spot price (New York Harbor) to the crude oil spot price (West Texas Intermediate, Cushing, OK) is 1.25 (before any GHG emission costs are taken into account). This assumption was based upon an analysis of DOE/EIA data on diesel fuel and crude oil for the period January 2002 through July 2008.

Diesel fuel consumed for on-road transportation is composed of ultra-low-sulfur diesel (ULSD) (<15ppm sulfur) and low-sulfur diesel (LSD) (15-500ppm sulfur). Monthly U.S.

consumption rates for ULSD and LSD were estimated using EIA data on U.S. diesel production and U.S. diesel exports/imports. Based on these consumption rates, the ULSD percentage of the total on-road diesel consumption was calculated for each month. An average of EIA retail price data showed that, on average, the monthly retail price for ULSD was 7.5 cents higher than the retail price for LSD during the period February 2007 to July 2008, and it was assumed that this difference also applied to their spot prices[52]. Using the monthly ULSD percentages and the 7.5 cent price difference, monthly ULSD spot prices were estimated from EIA's monthly spot prices for on-road diesel fuel (which aggregated ULSD with LSD). The ULSD/WTI price ratio was then calculated for each month during this period, and the average of the monthly ratios was 1.25.

A.5.3. Electric Power Price

If excess power is generated, it is assumed that the excess power is sold at a price of $50/MWh (Jan-2008 $). Note that this price did not factor heavily in this study. Only one case generated excess power: the CTL without CCS case generated 38 MW-net, which equates to about 1% of the total product slate energy value.

A.6. Escalation

All costs (e.g., fuel costs, O&M costs) and product prices (e.g., diesel, naphtha, and power prices) are assumed to escalate at an annual nominal rate of 2%. Capital costs are assumed to escalate at the same rate during the construction period. Credits or debits for CO_2 emissions are also assumed to escalate at the same rate.

The escalation rate was based on the GDP Chain-type Price Index reported in Table A19 of the EIA's Annual Energy Outlook 2008, which was projected to grow at an annual nominal rate of 2.0% between 2006 and 2030.

A.7. Depreciation, Taxes And Insurance

Depreciable financing and capital costs were depreciated over twenty years using the 150% declining-balance method.

An effective total income tax rate of 38% was assumed.

Annual property taxes and insurance costs were assumed to be equal to 2% of the total "overnight" capital cost (all capital costs excluding escalation and interest during construction).

A.8. References

[1] Worhach, P. and J. Haslbeck, "Recommended Project Finance Structures for the Economic Analysis of Fossil-Based Energy Projects," DOE/NETL-401/090808, National Energy Technology Laboratory, September 2008, http://www.netl.doe.gov/ energyanalyses/pubs/Project%20Finance%20Parameters%20- %20Final%20Report%20- %20Sept%202008 1.pdf.

[2] Higman, C., "Integrated Gasification Combined Cycle (IGCC) Design Considerations for High Availability – Volume 1: Lessons from Existing Operations," EPRI 1012226, March 2007, http://mydocs.epri.com/docs/public/000000000001012226.pdf.
[3] Ortiz, D.S. et al, "Characterization of Biomass Feedstocks," RAND Corporation PM-216-NETL, May 2007.
[4] Keamey, D., Energy Information Administration, Personal Communication, Unpublished reference case data from Annual Energy Outlook 2008 National Energy Modeling System runs AEO2008.D030208F and HP2008.D031808A, August 2008.
[5] Gray, D., Personal Communication, January 2008.

B. Greenhouse Gas Accounting Methodology

The greenhouse gas (GHG) accounting methodology used in this study is based on a life cycle assessment (LCA) approach. The general approach and allocation procedures are consistent with the guidelines for performing LCA's developed by the International Standards Organization (ISO). (ISO 2006, ISO 2006a)

The scope of this study is based on the production, delivery, and use of low-sulfur diesel fuel produced in coal-toliquids (CTL), coal/biomass-to-liquids (CBTL), and biomass-to-liquids (BTL) energy conversion facilities in the United States. Illinois No. 6 bituminous coal is used as the fossil energy feedstock and switchgrass is used as the biomass feedstock. The energy conversion facilities are modeled to represent conceptual, third- or fourth-of-a-kind, plant designs integrating coal and biomass gasification with Fischer-Tropsch liquids production, and carbon dioxide capture and storage (CCS). The time-period represented by the energy conversion facilities is considered a near-term technology based on the integration of commercially-available technology.

Comparative analysis of the life cycle greenhouse gas emissions to petroleum-derived diesel fuel is based on an industry size representing one million barrels per day (or more) production of FTD. No future energy efficiency or learning curves estimates are accounted for in the comparative assessment modeling approach. The petroleum-derived diesel fuel GHG estimate is based on year 2005 national average for conventional internal combustion engine passenger vehicle based on the NETL Life Cycle GHG Petroleum-based Fuels Model. Year 2005 petroleum baseline was selected as the comparative baseline year in conformance with the Energy Independence and Security Act of 2007 (EISA 2007).

The GHG accounting methodology was developed using industry standards to match the scope of this study and provide a comparative GHG methodology. The following describes the modeling approach and comparative petroleum-based diesel fuel baseline.

B.1. Study Boundary and Modeling Approach

The study boundary for the analysis is from the extraction and harvesting of raw materials from the earth to the consumption of the diesel fuel to move a passenger vehicle. The boundary applied is commonly referred to as a "cradle-to-grave" life cycle assessment. The "cradle" refers to extraction of raw materials from the earth and the "grave" is represented as the consumption of the fuel in the vehicle. Figure B-1 graphically represents the boundary of the life cycle operations included within this study. Within transportation studies (GM 2001) the boundary is also referred to as a "well-to-wheels (WTW)" analysis. The term "well" originates from studies analyzing conventional crude oil extracted from on-shore and off-shore wells for the production of transportation fuels. The term "wheel" refers to the use of the fuel to turn the wheels on the vehicle. The definition of "wellsto-wheels" is expanded within this study to encompass any energy material extracted or harvested (e.g., coal, biomass and crude oil) from the earth for the purposes of producing transportation related fuels.

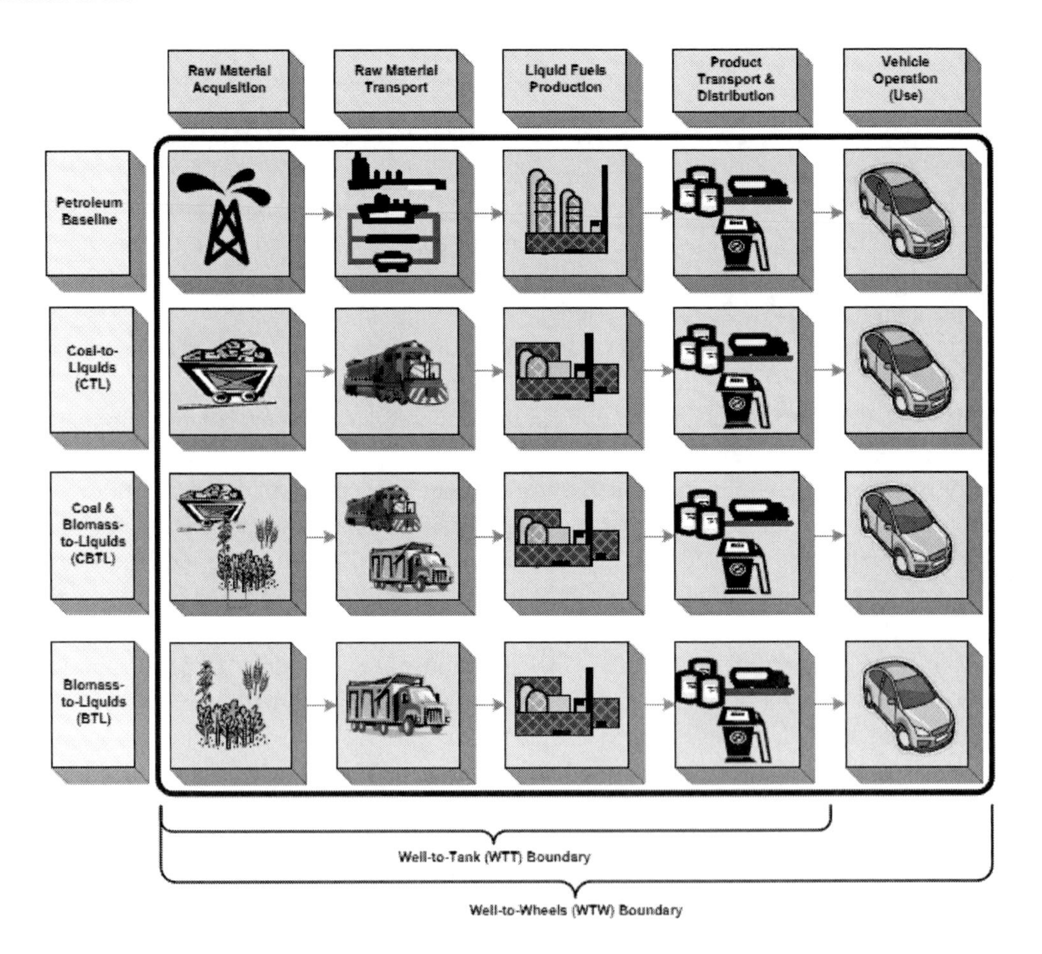

Figure B-1. Life Cycle Stages Included in the Study Boundary.

Contributions of GHG emissions are accounted for within each of the following life cycle stages:

- Raw Material Acquisition (e.g., coal mining and growing and harvesting of biomass).
- Raw Material Transport (e.g., rail transport of coal from the coal preparation plant to the coal-to-liquids plant, truck transport of biomass from the field collection point to the coal/biomass-to-liquids plant).
- Liquids Fuels Production (e.g., coal-to-liquids plant, coal/biomass-to-liquids plant, petroleum refinery). Carbon capture and storage (CCS) operations are included within this life cycle stage.

- Product Transport & Distribution (e.g., transport of the diesel fuel from the energy conversion plant to the refueling station, on-site storage, and dispensing of the fuel into a vehicle).
- Vehicle Operation (e.g., combustion of the fuel in a passenger vehicle).

B.1.1. Scope of the Environmental Life Cycle Analysis

The environmental LCA approach utilizes the International Standards Organization (ISO) 14040 "Environmental Management – Life Cycle Assessment – Principles and Framework." (ISO 2006) This study includes all four phases of a life cycle assessment: Goal and Scope Definition, Inventory Analysis, Impact Assessment, and Interpretation. The scope of the life cycle inventory (LCI) is limited to greenhouse gas emissions, as a result, the life cycle impact assessment (LCIA) only determines the global warming potential (GWP) of the GHG emissions based on their relative contribution.

The scope of the GHG emissions accounted for within the study is limited to carbon dioxide (CO_2), methane (CH_4), and nitrous oxide (N_2O). The total contributions of all other GHGs are deemed to contribute less than one percent to the total environmental impact. The effects of each GHG emission are normalized and reported in terms of their global warming potential (GWP). Normalized values are expressed in terms of CO_2 equivalents (CO_2E).

The Intergovernmental Panel on Climate Change (IPCC) publishes the international standard for calculating GWP based on the weighted contribution of various emissions (IPCC 2001, IPCC 2007). The IPCC publishes values for three time-frames: 20, 100, and 500 years. The U.S. standard is based on 100-year time-frame. GWP were standardized in 1990, 1996, 2001, and in 2007 by the IPCC. Within this study the 2007 IPCC values are used. Table B-1 lists the primary GHGs and their corresponding global warming potentials (GWP) reported in mass of CO_2 equivalents.

Table B-1. GHG Emissions Included in Study Boundary and their 100-year GWP

Emissions to Air	Abbreviation	2001 IPCC (GWP, CO_2E)	2007 IPCC (GWP, CO_2E)	This Study (GWP, CO_2E)
Carbon Dioxide	CO_2	1	1	1
Methane	CH_4	23	25	25
Nitrous Oxide	N_2O	296	298	298

Nitrogen oxides (NOx) and their impact to global warming are currently being reviewed by climatologists around the world. As a result, there is a lack of agreement about the impact of NOx in relation to global warming. The 2007 Intergovernmental Panel on Climate Change (IPCC) report entitled "Climate Change 2007: The Physical Science Basis" notes that nitrogen oxides have short lifetimes and complex nonlinear chemistry with opposing indirect effects through ozone enhancements and methane reduction (IPCC, 2007). Most current research suggests that the GWP for surface/industrial NOx emission may be negative. Wild et al. (2001) report a GWP for industrial NOx emissions of -12. Since there is a lack of agreement on NOx effects, the IPCC has opted to omit them from consideration.

B.1.2. Cut-off Criteria for the System Boundary

Cut-off criteria define the selection of materials and processes to be included in the system boundary. Following the requirements of ISO14040 (ISO 2006), the criteria of mass, cost, and environmental relevance was used for material and energy inputs.

A significant material input is defined as a material that has a mass greater than 0.01 lb per lb of the principal product that is produced by a unit process. A significant material is also one that has a relatively high cost (for instance, compared to the cost of the largest, by mass, material input), or has an important environmental relevance (for instance, a high global warming potential).

A significant energy input is defined as one that contributes more than 1% of the total energy used by the unit process. As with materials, a significant energy input is also one that has a relatively high cost or has an important environmental relevance.

B.1.3. Exclusion of Data from the System Boundary

All physical operations are considered pre-existing. Therefore, no construction related emissions are accounted for within the study. The production and delivery of electricity and other fossil-fuels to support each life cycle stage are accounted for within the study. For example, the GHG emissions for petroleum-derived diesel fuel used to transport biomass from the field collection point to the energy conversion facility are inclusive of the emissions associated with the extraction, refining, and transport of the fuel to the end-user (i.e., the truck used to transport the biomass). Contributions and/or credits of GHG emissions from land use change are not included within the study because all operations are considered pre-existing; therefore no changes in land use occur. With regards to biomass production, only so-called "marginal" lands are used, such that switchgrass production will not result in land-use change or crop displacement. Storage of carbon in select biomass root structures has also been excluded from the analysis. However, carbon reductions from soil root carbon storage may occur for select biomass types and could be accounted for in similar studies.

Humans involved in the system boundary have a burden on the environment, such as driving to and from work and production of food they eat, that is part of the overall life cycle. However, this complicates the life cycle tremendously due to the data collection required to quantify the human-related inflows and outflows on the environment and how to allocate them to fuel production. Furthermore, it is assumed that the workforce will be unaffected by the choice of fuel. Issues related to humans, such as the societal impacts of humans in the workforce that need to be addressed through policy and value-based decisions, are outside the scope of a life cycle study.

Low frequency, high magnitude environmental events (e.g., routine/fugitive/accidental releases) were not included in the system boundaries, since such circumstances are difficult to associate with a particular product.

More frequent, but perhaps lower magnitude events, such as material loss during transport, are included in the system boundary.

B.1.4. Data Resources and Analysis Tools

Secondary data sources were used to model each life cycle. A range of industry reported emissions data, publically-available literature data, and industry emission factors were used to characterize each life cycle stage. Raw material acquisition of coal and biomass was modeled by RAND Corporation (RAND) under contract to NETL. (Ortiz 2007) A detailed process model was developed by NETL and Noblis to model each energy conversion facility (CTL, CBTL, and BTL). Product transport and distribution of the fuel was modeled using the GREET Model, Version 1.8b and the EPA MOVES model was used to determine the carbon content of the fuel. Table B-2 summarizes the types of data used to characterize each life cycle stage, the primary analysis tool, and the scope of the life cycle stage.

Table B-2. Data Resources and Analysis Tools

Life Cycle Stage	Data Type	Analysis Tool	Scope of Analysis
Raw Material Acquisition	Industry Reported Emissions & Operating Data by Raw Material Type	RAND Coal & Biomass Model, Developed for NETL (2008)	Illinois No. 6 Bituminous Coal, Switchgrass, and Crude Oil
Raw Material Transport	Industry Reported Emissions & Operating Data by Raw Material Type	RAND Coal & Biomass Model, Developed for NETL (2008)	Coal - 200 miles by rail, Biomass - 40 - 70 miles (based on biomass feed rate) by heavy duty truck
Liquid Fuels Production	Industry Data	Aspen Process Models	5,000 to 50,000 barrels per day FTL plant size, including carbon sequestration where applicable
Product Transport & Distribution	GREET Emission Profiles	GREET v1.8b	GREET Model Year 2005, Default Value for FT Diesel
Vehicle Operation (Use)	EPA, Office of Transportation and Air Quality Modeling Estimates	U.S. EPA, MOVES Model	Diesel Powered, Conventional Compression Ignition Engine, Passenger Vehicle

B.1.5. Data Reduction and Allocation Procedures

In order to generate the life cycle inventory, collected secondary data was reduced using numerous calculations and equations. These reductions manipulated the secondary data to the goal and scope of this study (e.g., relating data to the functional unit, data aggregation, allocation of flows/releases, etc.). ISO 14044 Section 4.3.3 (calculating data) and 4.3.4 (allocation) standards, where appropriate, are used as guidelines in performing data reductions and allocation procedures (ISO 2006a).

System expansion is one of the two methods recommended within ISO 14044 for avoiding allocation wherever possible (i.e., avoiding allocation is preferred) and is used within this study. The displacement method, a type of system expansion, is recommended by the U.S. Environmental Protection Agency (EPA) for allocating co-products from energy conversion facilities producing transportation fuels. (EPA 2007) When avoiding allocation is determined not to be feasible, energy carriers or any material produced for its energy value was allocated based on energy content. All other materials and co-products are allocated based on mass when applicable.

Specifically, the displacement method is used in this study to determine the GHG emissions for the naphtha and electricity co-products produced by the CTL/CBTL facilities.[53] The displacement method expands the system boundary to include the production of co-products by other means that would theoretically be avoided as a result of secondary production by the primary process being modeled. For example, electricity is produced by various means within the U.S. and delivered to the electricity grid for distribution to end-users. Electricity produced as a co-product from a CTL or CBTL facility will also be delivered to the electricity grid for distribution to end-users. Therefore, an equivalent quantity of electricity would theoretically be displaced (not needed or off-set new generation) from the average U.S. electricity generation base. Using the displacement method, the U.S. average GHG profile for the generation and distribution of electricity (mass of CO_2E per unit of energy delivered) is determined to be equivalent to the co-product credit (or offset) for the electricity produced from the CTL or CBTL facility. This methodology is applied consistently throughout the study.

B.1.6. Basis of Comparison

The change in life cycle GHG emissions from FT diesel was determined within this study by comparison to a 2005 national average petroleum-derived diesel fuel baseline. Results are reported per million British Thermal Units (MMBtu) of net energy (fuel) consumed (i.e., on a lower heating value [LHV] basis). Comparing alternative sources of transportation fuel on a "per mmBtu LHV of fuel consumed" basis ensures equivalent vehicle efficiency profiles are used and improves comparability of upstream (well-to-tank) life cycle emissions.

B.2. Life Cycle GHG Accounting Procedure

The following equation summarizes the calculation procedure for determining the GHG emissions for FT diesel produced from a CTL or CBTL facility with co-products.

Total Cradle-to-Gate GHG Emissions for Producing Diesel, Naphtha, and Electricity (Raw Material Acquisition thru the exit gate of the energy conversion facility)

(−) Carbon Content (converted to CO_2) of the Biomass Feedstock Utilized by the Energy Conversion Facility (applicable to CBTL facilities only)

(−) Naphtha Cradle-to-Gate GHG Co-product Displacement Value per Million Btu of Naphtha Produced

(−) Electricity Cradle-to-Gate GHG Co-product Displacement Value per Million Btu of Electricity Produced

(+) GHG Emissions from Transportation & Distribution of Diesel Fuel to the End User

(+) GHG Emissions from Combustion of the Diesel Fuel in the End Users Vehicle

Total Life Cycle GHG Emissions for FT diesel Produced from a CTL or CBTL Facility

The methodology applied for determining the biomass carbon credit, naphtha co-product displacement value, and electricity co-product displacement value are summarized below. The life cycle stage results are reported with study results for each case analyzed.

B.2.1. GHG Emissions from Coal Mining and Transportation

The mining and transportation of coal results in two primary GHG emissions: (1) the release of methane adsorbed within the coal seam, and (2) the operation of fossil fuel powered from mining/transportation equipment. Methane emissions have the potential to represent a significant portion of the overall life cycle GHG emissions for coal based processes and are therefore dealt with here in some detail. The non-methane emissions were calculated using the GREET model (version 1.8a) using the heating value of Illinois #6 coal and assuming a transport distance of 200 miles by rail.

The formation of coal occurs by a process called coalification, in which biomass is converted to coal over the course of millions of years under high pressures and temperatures. Methane gas is formed as part of this process and a portion of this methane remains in the coal seam and surrounding rock strata as coalification continues.[54]

The methane content of different coals varies widely, across coal types, basins, and even within the same basin. During mining operations, this coal mine methane (CMM) is released, resulting in a GHG emission, but the exact emissions are dependent on the type of mining, methane content of the coal, and end use of the coal.

While these issues make the estimation of CMM emissions rates difficult, the EPA has identified 50 mines which result in 95% of emissions from underground coal mining operations. [REF: "Identifying Opportunities for Methane Recovery at U.S. Coal Mines: Profiles of Selected Gassy Underground Coal Mines 1999-2003", EPA Publication: EPA 430-K-04-003] This study uses the average emission rates from six of the seven gassy Illinois basin mines on this list as a conservative estimate of CMM emissions for Illinois #6 coal. One Illinois basin mine – the Baker mine which produces W. Kentucky #13 coal – was omitted from the list as it is an outlier in the emissions rates (well over three times the average emissions from the other six mines), making it by far the gassiest of mines in that basin.

Keeping with premise that carbon legislation has been passed (a key premise of this CBTL study), gassy mines are assumed to have implemented "best practice" CMM recovery methods to avoid GHG emissions and to supplement existing mine safety requirements.[55] The amount of recoverable methane was determined using the average emission rates from the gassy mines in the Illinois basin, the maximum recoverable CMM using selected methods, and the average in-situ methane contents for Illinois basin coal. [EPA 1999, EPA 2005, EPA 1995] This methane is assumed to have been combusted on-site, either flared or for use in mining equipment, such as ventilation air heaters. However, no GHG displacement credit is taken as the end use of the methane is unknown. Similarly, an economic benefit or detriment is no assessed in association with methane recovery, as this is beyond the scope of the study. Instead, it is assumed that recovered methane pays for itself, if not in sales revenues, then in reduced operating costs as it is used as a fuel.

Table B-3. Upstream GHG Emissions from Coal Mining and Transportation

Emission Type	Emissions Rate	Units
Non-Methane Emissions from Mining	1.941	lb CO_2E/mmBtu Coal (LHV)
CO_2 Emissions from Rail Transportation	0.077	lb CO_2E/short ton coal/mile
Methane Emissions after CMM Recovery	96	lb CO_2E /short ton coal
Total Upstream Emissions with CMM Recovery	171	lb CO_2E /short ton coal

B.2.2. GHG Emissions from Biomass Cultivation, Harvesting, Processing, and Transportation

For agriculture, it is assumed that greenhouse gas emissions are predominantly the result of energy consumption and fertilizer use. The total primary farm-to-plant-gate greenhouse gas emissions are a sum of the carbon dioxide from cultivation and harvesting, transporting, and processing of biomass.

Cultivation and harvesting includes the steps of planting, fertilizing, cutting, and harvesting crops in the fields. Transportation captures emissions associated with moving crops from the field to the gasification facility. Processing accounts for emissions associated with any cutting, drying, grinding, densification or other steps necessary to transform crops so that they meet the volume, size, and moisture constraints for transportation, storage, and gasification processes used.

Estimates for biomass life cycle GHG emissions used in this study were based on a Coal and Biomass Model developed for NETL by RAND (RAND 2008). The emissions for each life cycle stage of switchgrass cultivation are summarized in Table B-4 and are based on an average feed rate of 4,000 short tons of switchgrass per day, or 1.3 million short tons per year delivered the plant gate.

Table B-4. Upstream GHG Emissions from Switchgrass Production Equivalent to 4,000 dry tons

Emission Category	Emissions Rate	Units
Planting and Harvesting	13,940	tons carbon/year
Agrichemicals Production	36,930	tons carbon/year
Transport of Agrichemicals	280	tons carbon/year
Production Sub-Total	51,150	tons carbon/year
Processing (Cutting, Sizing, Drying)	3,800	tons carbon/year
Transportation	3,240	tons carbon/year
Total Upstream Emissions with CMM Recovery	58,190	tons carbon/year

B.2.3. Biomass Carbon Credit

Carbon dioxide is extracted from the atmosphere by green plants and combined with sunlight and water to produce organic compound and oxygen; a process called photosynthesis. The carbon removed from the atmosphere by the plant is stored within the biomass structure of the plant. When the plant, in this case switchgrass, is harvested and gasified in a CBTL or BTL plant the carbon content of the biomass source is partially retained in the fuel, captured and sequestered, and released to the atmosphere through the CBTL/BTL stack. When the biomass carbon contained in the fuel is combusted in the vehicle it is released back into the atmosphere. For the purposes of this study, a conservative assumption that 100% of the carbon content of the diesel fuel combusted in the vehicle is converted to CO_2. A plant designed without carbon capture results in 100% of the carbon contained in the biomass feedstock eventually being returned back to the atmosphere; creating a net zero GHG effect. When CO_2 is captured and sequestered from a CBTL or CTL plant a portion of the carbon from the original biomass feedstock is captured and sequestered. This creates a biomass carbon credit because less CO_2 is returned to the atmosphere then was extracted by the green plant during photosynthesis.

A carbon credit is applied to the CBTL and BTL plants with CCS in this study based on the percent CO_2 captured and permanently sequestered.

B.2.4. Naphtha Co-product Displacement Value

The naphtha co-product displacement value presented in Table B-5 is based on the GHG profile for production of petroleum-based kerosene-type jet fuel at U.S. refineries in 2005 (Skone 2008). The profile is well-to-gate (raw material acquisition through exit of the liquid fuel production facility) and is presented on a per-barrel-refined basis. While kerosene has a higher boiling point range than naphtha, it is best suited to represent the displacement emissions profile for FT naphtha as both are primarily paraffinic (straight chain hydrocarbons). The modeled kerosene production process includes minimal upgrading and hydrotreating to reduce sulfur content. Gasoline and naphtha used as a petrochemical feedstock were not considered appropriate surrogates for FT naphtha because, while they have a similar boiling point range, both have additional upgrading requirements (and corresponding energy input and GHG emissions) to meet product specifications.

Table B-5. Naphtha Co-product Displacement Value

Co-product	Co-product Displacement Value (well-to-gate)	Units
Naphtha, as Fuel (petroleum-derived)	75.8	kg CO_2E/bbl refined product

B.2.5. Electricity Co-product Displacement Value

Certain emissions from electricity generating facilities are tracked by the U.S. Environmental Protection Agency (EPA 2007) and are publically available in the Emissions & Generation Resource Integrated Database (eGRID). While this database includes comprehensive coverage of CO_2 emissions from these generating facilities, it does not address other GHG constituents such as N2O and CH4 in its inventory.

The emissions data within eGRID only includes that attributable to operations and does not represent construction or upstream emissions. Therefore, U.S. average life cycle inventory profiles for upstream GHG emissions (raw material acquisition and transport) were added to the eGRID national data. Upstream life cycle inventory data sets were obtained from the

GaBi 4.0 Life Cycle Assessment Software – Professional Database – developed by PE International. (GaBi 4 2007).

A U.S. electric grid mix is representative of the year 2004 eGRID electricity mix (current update as of August 2008). Figure B-2 shows the source mix as a percentage of total U.S. electricity generation.

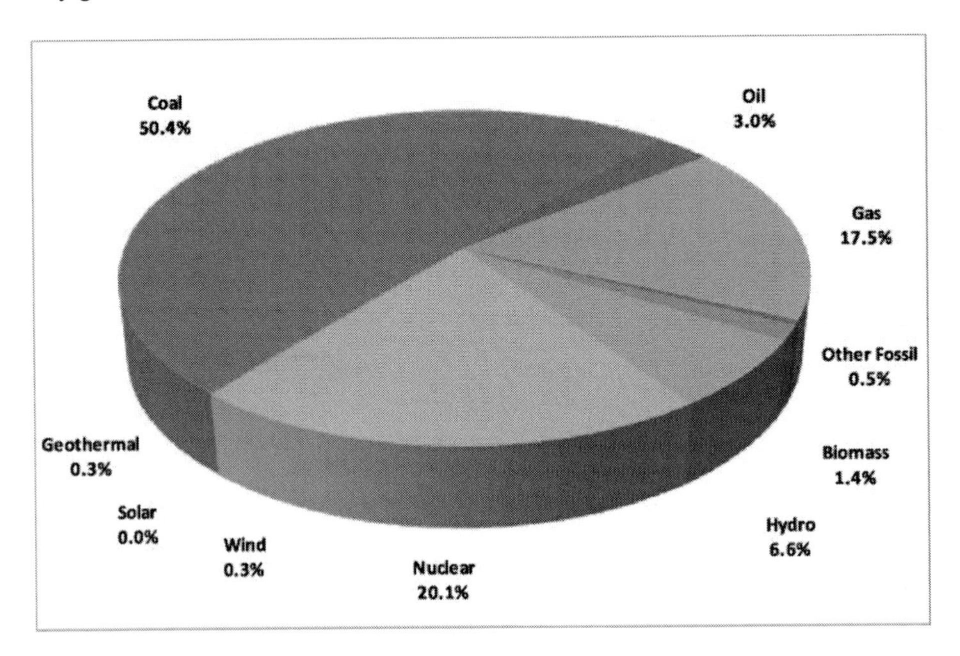

Figure B-2. Year 2004 Electricity Sources for the Average U.S. Grid Mix (EPA 2007).

Table B-6 summarizes the GHG profile used to determine the amount of GHG emissions subtracted for each unit of electricity co-product produced by the CTL and CBTL facilities.

Table B-6. Electricity Co-product Displacement Value

Co-product	Cradle-to-End User (CTEU)	Transportation & Distribution (T&D)	Co-product Displacement Value (CTEU minus T&D)	Units
Electricity (U.S avg. electricity grid mix, Yr. 2004)	223.7	8% Loss (15.7)	208.0	kg CO_2E/mmBtu

The petroleum-derived diesel fuel baseline used in this study was obtained from the DOE, NETL report entitled "Development of Baseline Data and Analysis of Life Cycle Greenhouse Gas Emissions of Petroleum-Based Fuels" dated November 2008. (Skone 2008) The baseline represents diesel fuel sold or distributed in the United States in the year 2005. The study goals and scope were aligned to meet the definition of "baseline lifecycle greenhouse gas emissions" as defined in the Energy Independence and Security Act of 2007 (EISA 2007), Title II, Subtitle A, Sec. 201.

The physical boundaries of the life cycle include operations that have a significant contribution to the total life cycle GHG emissions. Specifically, the average life cycle GHG profile for transportation fuels sold or distributed in the United States in 2005 is determined based on the weighted average of fuels produced in the U.S. plus fuels imported into the U.S. minus fuels produced in the U.S. but exported to other countries for use.

GHG life cycle results are reported in terms of kg CO_2E/MMBtu LHV of fuel consumed. This metric is dependent on the energy content of the fuel and vehicle efficiency and could alternatively be reported in terms of kg CO_2E/mile traveled. Table B-7 summarizes the life cycle GHG analysis study design parameters for conventional diesel fuel. Figure B-3 summarizes the life cycle GHG baseline emissions for petroleum-derived diesel fuel. The petroleum baseline for conventional diesel fuel sold or distributed in the United States in 2005, on a national average basis, is 95.0 kg CO_2E per million Btu, LHV, of fuel consumed.

Table B-7. Life Cycle Greenhouse Gas Study Desi

Life Cycle Boundary	Well-to-Wheels/Wake (Raw Material Extraction thru Fuel Use)
Temporal Representation	Year 2005
Technological Representation	Industry Average
Geographical Representation	Transportation Fuel Sold or Distributed in the United States
Impact Assessment Methodology	Global Warming Potential, IPCC 2007, 100-year time-frame
Reporting Metric	kg CO_2E/MMBtu LHV of Fuel Consumed
Data Quality Objectives	100% Publically Available Data Full Transparency of Modeling Approach and Data Sources Accounting for 99% of Mass and Energy Accounting for 99% of Environmental Relevance Process-based ("Bottoms-up") Modeling Approach

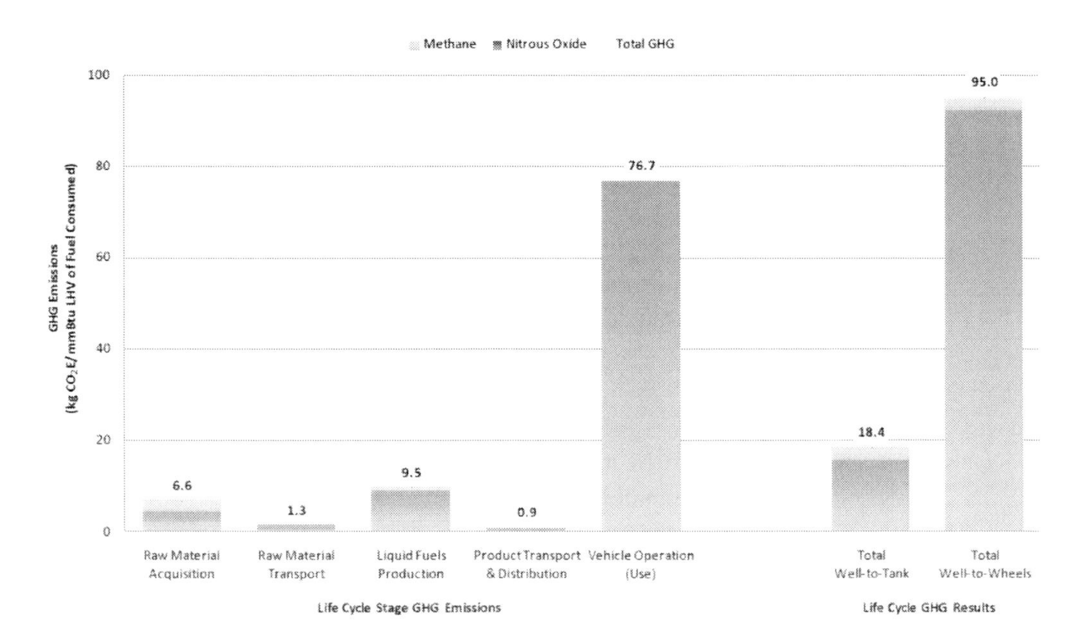

Figure B-3. Life Cycle GHG 2005 Baseline Emissions for Petroleum-derived Diesel Fuel (kg CO_2E/mmBtu LHV of Fuel Consumed).

B.4. References

EPA 2007

U.S. Environmental Protection Agency. Emissions & Generation Resource Integrated Database (eGRID). Updated (April 30, 2007): eGRID2006 Version 2.1. http://www.epa.gov/cleanenergy/energy

EPA 1999

http://yosemite.epa.gov/OAR/globalwarming.nsf/UniqueKeyLookup/SHSU5BUT5X/$File/methane emissions.pdf

EPA 1995

http://www.epa.gov/ttn/chief/ap42/ch14/related/mine.pdf

EPA 2005

http://www.epa.gov/cmop/docs/profiles 2003 final.pdf

GaBi 4 2007

GaBi 4.0 Life Cycle Assessment Software – Professional Database. PE International GmbH. Germany. [extracted from Gabi 4 software and data base for Life Cycle Engineering] 2007

Ortiz 2007

Ortiz, D.S. et. al. "Characterization of Biomass Feedstocks." RAND Corporation PM-2316-NETL. May 2007.

Skone 2008

Skone, T.; Gerdes, K. 2008. "Development of Baseline Data and Analysis of Life Cycle Greenhouse Gas Emissions of Petroleum-Based Fuels" November 2008. Department of Energy, National Energy Technology Laboratory, Pittsburgh, PA

C. Switchgrass and Coal Characterization

Feedstock flexibility of gasification-based systems offers the opportunity to take advantage of the benefits of domestically available and renewable biomass resources. The co-gasification of biomass with coal allows for the utilization of a wide range of biomass feedstocks, including residues from current agricultural activities (e.g., corn stover, forest residues) as well as energy crops that can be grown on marginal lands (e.g., mixed prairie grass, switchgrass, and short rotation woody crops).

A focus is placed on the use of a dedicated energy crop (e.g. switchgrass) cultivated on marginal lands as a feedstock for gasification or co-gasification. This focus means that no significant land use changes were anticipated based on the demand for biomass for a FT fuels industry. The use of other biomass feedstocks (e.g. mixed prairie grasses, corn stover, or woody biomass) was not anticipated to significantly change the results of the study with regards to the performance or economics of the CBTL plant configurations evaluated.

C.1. Biomass Characterisitcs

A study was performed to characterize potential biomass feedstocks for co-gasification in a CBTL facility.[56] The study characterized potential biomass feedstocks in respect to their chemical characteristics, regional resource quantities, delivered feedstock costs, and emissions of greenhouse gases during cultivation, harvest, and transportation to the plant gate. This study focused on the availability of marginal lands for dedicated energy crops (e.g., switchgrass, mixed prairie grass, SWRC) and existing land use practices for residues (e.g., corn stover, forest residues). This focus means that no significant land use changes were anticipated based on the demand for biomass for a FT fuels industry.

The report focuses on one of the five types of biomass resources were characterized in detail, switchgrass. Switchgrass is a perennial grass native to the United States. While some strains are better suited to different regions, switchgrass stands are drought resistant and prevent soil erosion. Switchgrass stands may also fix carbon in depleted soils.

C.1.1. Chemical Characteristics

Chemical composition and concentration of trace elements of various biomass feedstocks can vary based on a variety of ambient and local conditions. Information obtained from the open literature provided a range of values for switchgrass and these ranges are presented in Tables C-1 and C-2. Because the actual composition of the biomass types can vary over a wide range, these values represent an average composition.

Table C-1. Switchgrass Analysis

	Switchgrass Dry Basis %
C	46.96
H	5.72
O	40.18
N	0.86
S	0.09
Cℓ	0.00
Ash	6.19
Moisture	0.00
Total	100.00
HHV (MMBtu/dry ton)	16.12
HHV (Btu/dry lb)	8,060
Na2O (% composition of ash)	0.55
K2O (% composition of ash)	13.88
Ash Fusion Temperature	1016 °C

Table C-2. Biomass Moisture Contents

Moisture Content, % (by weight)	Switchgrass
At Harvest	15%
After Passive Field Drying	15%
After Processing, Sizing, and Drying at CBTL Plant	10%

Presence of trace elements and potentially toxic compounds (e.g., chromium and lead) can vary significantly due to proximity to industrial emissions sources. One source reviewed during the study provided trace element samples from red oak and loblolly pine residues that were harvested a mile from an abandoned smelter. The presence of trace elements of concern will be site specific and should be considered in the design process of the CBTL facility, including the emissions profile. It should be noted that the process considered for a CBTL application provides multiple opportunities to capture and remove trace elements from coal (e.g., activated carbon for mercury removal). The capture and removal of trace elements other than mercury, including cadmium and lead, could potentially be accommodated at a reasonable cost, and the co-firing of biomass in a CBTL system could provide an opportunity to reclaim contaminated soil. NETL is currently funding research directed at developing low-cost methods of capturing and removing trace elements from syngas.

C.2. Delivered Biomass Costs

The study provided cost estimates for delivered biomass to the plant gate. Figure C-1 provides estimated costs (January 2008$) of switchgrass as a function of biomass feed rate. These are the costs associated with crop establishment, cultivation, harvesting and storage, and transportation. Biomass is assumed to be field dried, then actively dried and baled, and stored in the field until such time when it is needed. Final feedstock processing is assumed to occur at the CBTL facility and is not included in the delivered cost estimates. Delivered feedstock costs in Figure C-1 were developed on a $/dry ton basis (January 2008$) (even though the biomass is still "wet" when delivered to the plant). These costs are directly influenced by crop yield. Individual biomass feedstocks were characterized in respect to gross and net crop yields to account for losses resulting from harvest losses and storage and handling losses.

It should be reiterated that these costs are based on the use of marginal lands for energy crops and current land use practices for residues. Therefore, regional crop density and cultivation proximity to the CBTL facility is not optimized (i.e., the CBTL facility is not considered to be in the center of a high crop density operation located to serve the CBTL facility). These costs were developed to estimate the likely delivered costs for biomass at an "early deployment" facility located in an existing region with suitable existing resources to support the delivery of sufficient biomass feedstocks. A combined industrial-agricultural facility optimized for biomass production could potentially result in delivered costs lower than those presented in Figure C-1.

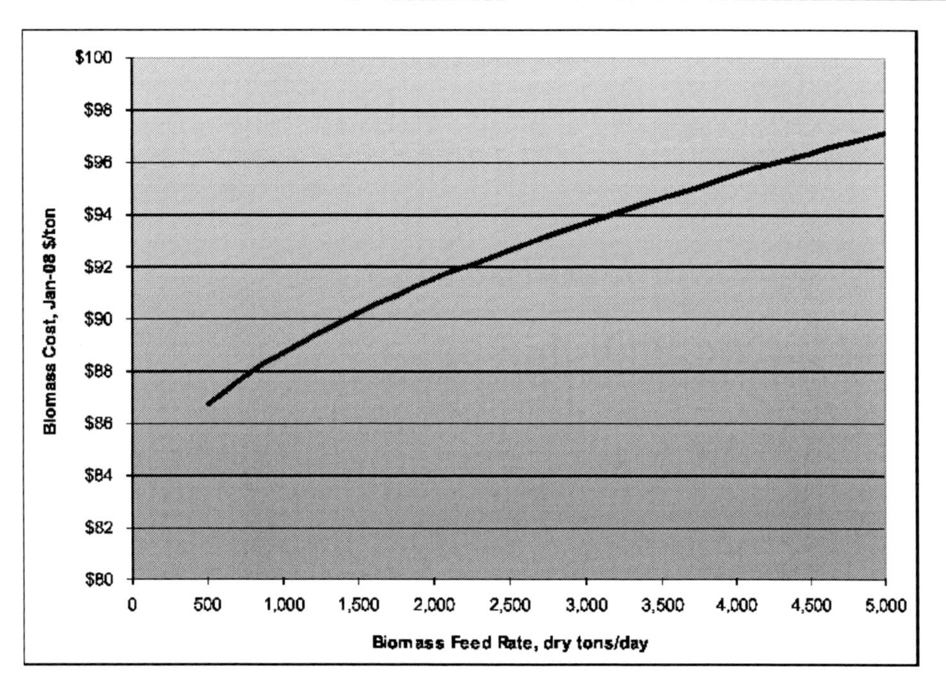

Figure C-1. Switchgrass Cost as a Function of Feed Rate.

C.2.1. Biomass Cost Components

A brief overview of major contributions to each cost component is included below.

- Establishment - Establishment of new plantings of switchgrass and mixed prairie grasses as dedicated energy crops are estimated to result in harvesting in the third year. Costs associated with the new plantings include farm related capital, variable and fixed operating cost, and land rent. Switchgrass plantings are assumed viable for ten years, after which a new planting would be established.
- Cultivation - Cultivation costs for switchgrass includes continued land rent as well as operating costs (fertilizer, weed and insect control, maintenance and labor).
- Harvesting and Storage - Costs associated with harvesting and storage include appropriate activities (e.g., mowing or cutting and chipping, raking and baling or stacking and piling), and farm-site storage prior to transportation to the CBTL facility.
- Transportation - Costs associated with transportation from the collection site to the plant gate include capital for dedicated equipment (e.g., truck tractor, flatbed trailer, chip van, bale handler), operating, and maintenance costs. Transportation costs are influenced significantly by mean travel distance to supply a given biomass feedstock delivery rate (e.g., 4,000 dry tons per day). Mean travel distance is dependent on biomass yield as well as area being harvested. Because this study focused on marginal land availability for dedicated energy crops (e.g., switchgrass), the area being harvested, including regional crop density (i.e., fraction of land cultivated) was estimated based on results of the regional resource availability of each feedstock characterized.

D. COST AND PERFORMANCE TABLES

This appendix summarizes the technical, environmental and economic performance of the eleven CBTL plants evaluated.

D.1. Technical and Environmental Plant Performance

The key metrics in comparing the cases are Required Selling Price (economic), Plant Efficiency (technical) and Life Cycle GHG Emissions Profile of the Fuel (environmental). These metrics provide insights into plant capital and operating costs, as well as how the fuels produced compare both environmentally and cost-wise to petroleum-derived fuels.[57] Note that all of the Required Selling Prices listed in this appendix do not include the effects of a carbon price on the fuels, an analysis of which was provided in Chapter 4.

D.1.1. Technical and Environmental Plant Performance Results

The performance of select plants are summarized in Table D-1. "Coal Feed" lists the feed rate of Illinois #6 coal in as-received short tons per day (tpd) at 100% capacity factor. "Biomass Feed" lists the feed rate of switchgrass in as-received tpd and 100% capacity factor.

"Biomass Mass %" and "Biomass Energy % (HHV)" connote the percentage of the feedstock which is biomass, on a mass and energy basis, respectively.

"Total Fuel (BPD)" connotes the amount of product fuels produced in terms of barrels of fuel produced per day (bpd). This product is split into 70% diesel fuel and 30% naphtha on a volumetric basis. A 50,000 bpd plant size was the basis for this analysis, although in the BTL and 30wt% CBTL cases, biomass availability limits plant size to 5,000 bpd and 30,000 bpd, respectively.

"Gasifier Trains" details the number of gasifier trains needed to produce the required amount of syngas to meet the specified fuels production capacity. All cases include a spare gasifier train in order to ensure plant availability, as this can have a significant effect on the economic viability of the process. As described in Chapter 1, the gasifiers utilized in the BTL cases are circulating fluidized bed (CFB) gasifiers, whereas the remaining cases utilize dry-feed, entrained flow gasifiers.

"FT Reactors" details the number of FT reactors required for each case. The FT reactors are identical for all cases and are sized at 2,500 bpd.

"Internal Power (MW)" and "Export Power (MW)" describe the amount of electricity generated by the plant.

Internal Power is used to power equipment in the plant, such as the Air Separation Unit (ASU). Export Power is the power generated in excess of that required for the plant's needs and is sold as a product. The plants were configured for maximum fuels production and any Export Power generated is the result of recycle loop limitations.[58] This only applies to the Case 1, CTL without CCS.

Table D-1. Overall Performance for CBTL Plants

	Case 1	Case 2	Case 3	Case 4	Case 5	Case 6	Case 7	Case 8	Case 9	Case 10	Case 11
Plant Description	CTL w/o CCS	CTL w/ CCS	CTL w/ CCS + ATR	7.7 wt% CBTL w/ CCS	15wt% CBTL w/CCS	15wt% CBTL w/ CCS + ATR	30wt% CBTL w/ CCS	30wt% CBTL w/ CCS + ATR	BTL w/o CCS	BTL w/ CCS	BTL w/ CCS+ ATR
Coal Feed (as-received TPD, 100% CF)	21,719	21,214	23,035	19,948	18,923	20,667	9,893	10,843	0	0	0
Biomass Feed(as-received TPD, 100% CF)	0	0	0	1,657	3,339	3,647	4,240	4,647	4,084	4,136	4,350
Biomass Mass %	0.0%	0.0%	0.0%	7.7%	15.0%	15.0%	30.0%	30.0%	100.0%	100.0%	100.0%
Biomass Energy % (HHV)	0.0%	0.0%	0.0%	4.9%	9.9%	9.9%	21.0%	21.0%	100.0%	100.0%	100.0%
Total Liquids(BPD)	50,000	50,000	50,000	50,000	50,000	50,000	30,000	30,000	5,000	5,000	5,000
Diesel(BPD)	34,253	34,270	34,296	34,292	34,292	34,295	20,575	20,575	3,425	3,434	3,431
Naphtha (BPD)	15,747	15,730	15,704	15,708	15,708	15,705	9,425	9,425	1,575	1,566	1,569
Gasifier	11 train, entrained	11 train, entrained	11 train, entrained	11 train, entrained	11 train, entrained	12 train, entrained	8 train, entrained	8 train, entrained	6 train, CFB	6 train, CFB	6 train, CFB
FT Reactor	10 trains	10 trains	10 trains	10 trains	10 trains	10 trains	6 trains	6 trains	2 trains	2 trains	2 trains
Internal Power (MW)	415.6	450.7	544.8	447.8	451.8	550.3	272.2	333.4	60.2	69.0	78.7
Export Power (MW)	35.2	0.0	0.0	0	0	0	0	0	0.0	0.0	0.0
HHV Efficiency(%)	52.4	53.0	48.8	53.6	53.6	49.0	53.9	49.1	44.3	43.7	41.6
CO_2 Capture (%)	n/a	91.0%	96.6%	91.6%	91.3%	96.5%	91.5%	96.7%	n/a	87.5%	95.6%
Carbon Sequestered(tpd CO_2eq, 100% CF)	n/a	26,646	32,248	26,470	26,646	32,402	15,983	19,576	n/a	3,821	4,503
LCA Effective Carbon(gCO_2E/MMBtuLHV)	**235,000**	**90,200**	**83,700**	**76,000**	**63,400**	**55,300**	**35,100**	**23,800**	**-8,760**	**-210,430**	**-244,800**
LCA Carbon –Comparison to Petroleum Diesel	**+147%**	**-5%**	**-12%**	**-20%**	**-33%**	**-42%**	**-63%**	**-75%**	**-109%**	**-322%**	**-358%**

[1] The coal used in all cases is a bituminous, high-sulfur coal from the Illinois Basin: Illinois #6.

[2] The biomass used in all cases is Switchgrass.

"HHV Efficiency (%)" describes the overall thermal efficiency of the plant. This is the metric to be used in comparing the performance of the different plant configurations, regardless of plant output.[59] The efficiency also has a direct bearing on the relative plant size and fuel costs of the plant, and the latter constitutes the significant portion of Variable Operating Costs and totally O&M costs, as mentioned above.

"CO_2 Capture %" is the percentage of the CO_2 produced by the plant which is captured for sequestration (instead of being emitted). "LCA Effective Carbon (gCO_2e/MMBtu)" describes the limited-life cycle carbon content of the diesel fuel product in terms of grams of CO_2 equivalents per MMBtu of product (LHV basis). "LCA Carbon – Comparison to Petroleum-Derived Diesel" describes how the LCA Effective Carbon compares to that of petroleum-derived diesel. This is the percentage increase or decrease in the life cycle emissions of the fuel compared to petroleum-derived diesel.

D.2. Economic Results

Table D-2 provides an economic summary of the select plants described above in the plant performance section.

D.2.1. Capital Cost

The required capital investment for plants is commonly reported in terms of dollars per daily barrel produced ($/DB), and is listed under "Capital Requirement ($/DB)" in Table D-2. This is the overnight installed cost of the plant. Required capital investment is strongly dependent on the plant size, plant efficiency (less efficient plants require more fuel through them, and therefore larger equipment), whether the plant is equipped for CCS, and whether biomass is used as a feedstock or not.

D.2.1.1. Capital Cost Methodology

Equipment costs were estimated for all of the unit operations in the plants. These were then aggregated into subtotals for each major plant section:

- Coal and biomass handling, preparation and feeding - includes all equipment associated with the storage, reclaiming, conveying, crushing, preparation, drying, and sampling of coal and biomass feeds.
- Water systems - includes the cooling water systems, boiler feedwater systems, waste water treatment and other plant water treatment systems.
- Gasification - includes coal and biomass feed systems, the gasifiers, quench system, and slag removal.
- Air separation - includes a standard cryogenic system for separation of oxygen and nitrogen.
- Syngas cleaning and shift - includes several components that remove hydrogen sulfide, carbonyl sulfide, cyanide, ammonia, particulates, mercury, and carbon dioxide. It also includes acid gas treatment to remove hydrogen sulfide and bulk removal of carbon dioxide, sulfur recovery, hydrogen recovery and water gas shift.

- Carbon dioxide removal and compression - includes the amine system for removal of the carbon dioxide in the FT recycle loop, dehydration, and compression of the carbon dioxide to 2,200 psi.
- Carbon dioxide transport storage and monitoring - includes the transportation of CO_2 via pipeline, injection and storage into a saline geologic formation and monitoring for 80 years.
- FT synthesis - includes the FT slurry phase synthesis reactors, catalyst activation, FT product upgrading that includes wax hydrocracking, hydrotreating and product distillation, and hydrocarbon recovery.
- Power block - includes a boiler for production of high pressure steam and a steam turbine.
- Balance of plant - includes product tankage, ash handling, the accessory electrical plant, instrumentation and controls, site improvements, and buildings and structures.

The sum of the major plant section costs equals the Bare Erected Cost (BEC) for the plant. The following costs are added to the BEC to yield the total overnight capital requirement.

- Home office cost - Estimated as 9% of BEC. Includes costs for detailed design (including the Front-End Engineering Design study), construction permitting (but not environmental permitting), construction management and architect-engineer support during startup.
- Process contingencies - Added to compensate for uncertainty in cost estimates caused by performance uncertainties associated with the development status of a technology. Process contingencies were assigned to each major plant section based on engineering judgment and consideration of how much engineering data was used to develop the models for that section).
- Project contingency – Equal to 15% of the sum of BEC and process contingencies. Added to compensate for uncertainty in cost estimates caused by incomplete technical definition for these non-site specific feasibility analyses.
- Inventory cost (non-depreciable) – Includes the cost of spare parts, stored feedstocks (fuels and other consumables stored on-site), and first fills of chemicals and catalysts within process plant vessels. Estimated to be the sum of: a) one month of fuel costs at 100% capacity factor, b) one month of non-fuel variable operating and maintenance costs at 100% capacity factor, b) the cost of first fills of chemicals and catalysts, and c) the cost of spare parts, estimated at 0.5% of the sum of BEC and home office costs.

Table D-2. CBTL Plant Economic Summary

	Case 1	Case 2	Case 3	Case 4	Case 5	Case 6	Case 7	Case 8	Case 9	Case 10	Case 11
Plant Description	CTL w/o CCS	CTL w/CCS	CTL w/CCS + ATR	7.7wt% CBTL w/ CCS	15wt% CBTL w/ CCS	15wt% CBTL w/ CCS + ATR	30wt% CBTL w/ CCS	30wt% CBTL w/ CCS + ATR	BTL w/oCCS	BTL w/CCS	BTL w/CCS +ATR
Total Overnight Required Capital($/DB)	102,500	106,200	113,100	114,000	114,700	124,800	129,300	138,800	217,200	229,600	237,000
Fixed O&M ($MM/yr)	256	263	280	282	283	306	194	206	58	61	62
Non-Fuel Variable O&M ($MM/yr)	54	53	58	56	56	59	34	36	7	7	7
Biomass ($MM/yr)	0	0	0	44	92	102	119	132	108	110	116
Coal ($MM/yr)	308	301	327	283	268	293	140	154	0	140	268
Power Credit ($MM/yr)	29	0	0	0	0	0	0	0	0	0	0
Capacity Factor	90%	90%	90%	85%	90%	90%	90%	90%	85%	85%	85%
RSP Diesel ($/gal pet-diesel equiv.)	$2.49	$2.56	$2.74	$2.75	$2.82	$3.07	$3.23	$3.46	$6.44	$6.71	$6.95
RSP Crude Oil Equivalent ($/bbl)	**$84**	**$86**	**$92**	**$92**	**$95**	**$117**	**$109**	**$132**	**$216**	**$225**	**$234**

- Preproduction cost (non-depreciable) – Includes operator training, equipment checkout and startup costs. Estimated to be the sum of: a) one year of operating labor, b) one month of fuel costs at 100% capacity factor, and c) one month of non-fuel operating and maintenance costs at 100% capacity factor.
- Owner's cost (non-depreciable) - Includes: a) land cost ($3,000/acre), b) transmission interconnection cost, assumed to be $50 million for all cases, c) financing fees, estimated at 2% of total overnight required capital, and d) legal and environmental permitting costs, assumed to be $10 million for all cases.
- CO_2 monitoring (non-depreciable) – Includes: a) the initial capital cost for monitoring at the sequestration field, and b) O&M costs for monitoring the sequestration field over a period of eighty years (a thirty-year injection period followed by fifty years of post-injection monitoring). Estimated at $0.176 per metric ton of CO_2 stored (cumulative) over the project's economic life (30 years).

D.2.2. Operating Costs and Methodology

"Variable O&M ($MM/yr)" and "Fixed O&M ($MM/yr)" represents the annual operating expenses while "Biomass ($MM/yr)" and "Coal ($MM/yr)" represent the costs of the coal and biomass feedstocks. The "Power Credit ($MM/yr)" is the revenue stream generated from power sales. These values are all reported in millions of dollars per year and based on the plant operating at a 90% capacity factor (cases where coal is used) or 85% capacity factor (biomass only cases).

Fixed operating costs include pipeline and CO_2 storage maintenance, royalties, labor costs, overhead costs, administrative costs, local taxes, insurance, and maintenance materials. Non-Fuel Variable operating costs include the costs of: catalysts and chemicals, water and solids disposal. The small quantities of natural gas required for startup are not included.

D.2.3. Economic Results and Methodology

The required selling price (RSP) for the diesel fuel is calculated using the methodology in Chapter 4, yielding the "RSP Diesel ($/gal pet-diesel equiv)". "RSP Crude Oil Equivalent ($/bbl)" is the crude oil price that would yield a petroleum-derived diesel price that is equivalent to the "RSP Diesel".

D.3. Process Flow Diagrams and Stream Tables

The following section contains simplified process flow diagrams (PFDs) of five select cases in this report:

- CTL without CCS (Case 1)
- CTL with CCS (Case 2)
- CTL with CCS+ATR (Case 3)
- 15wt% CBTL with CCS (Case 5)
- BTL without CCS (Case 9)
- BTL with CCS (Case 10)

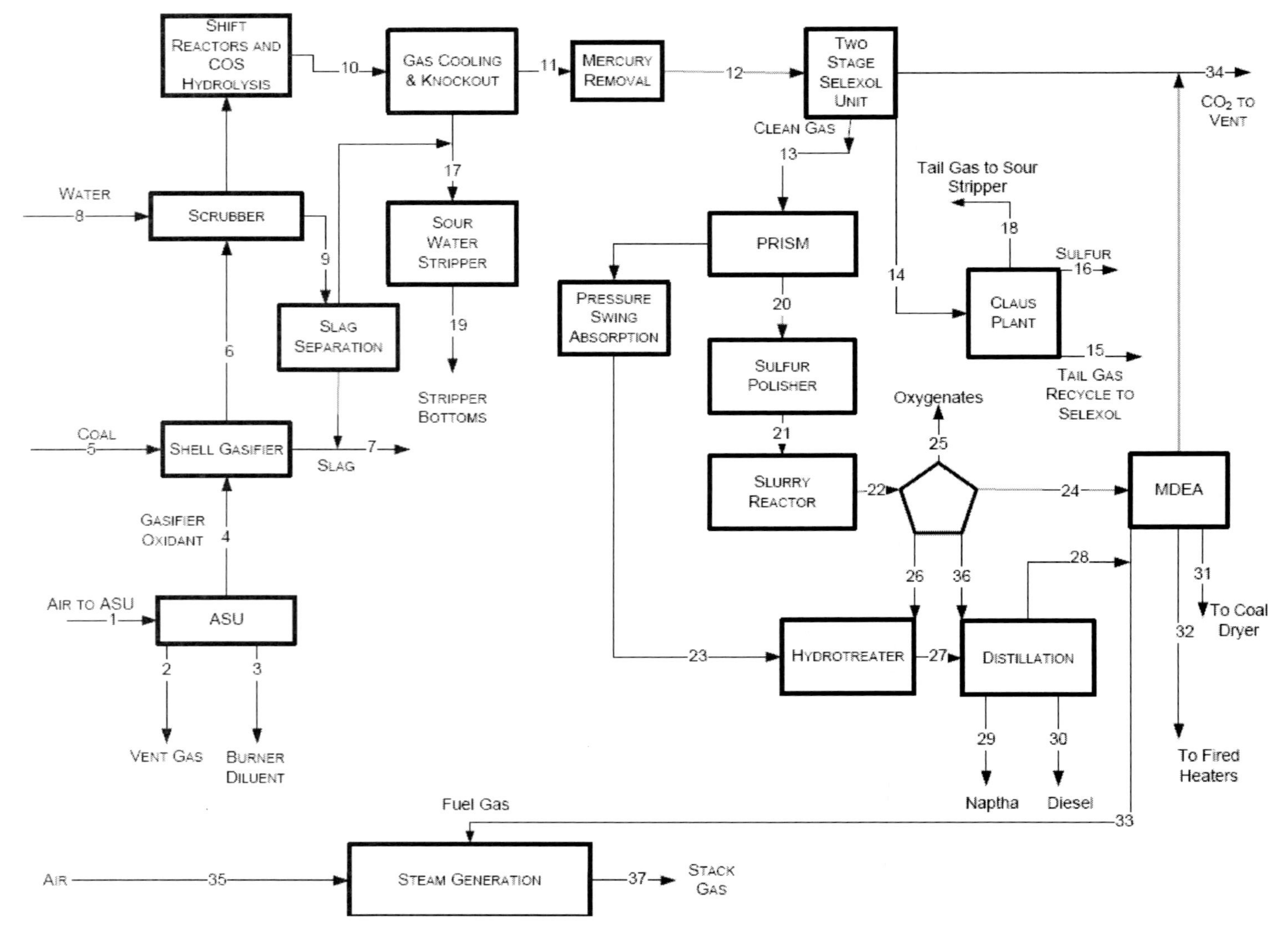

Figure D-1. CTL without CCS (Case 1).

Table D-3. Case 1 Stream Table

Stream #	1	2	3	4	5	6	7	8	9	10	11	12	13	14	15	16	17	18	19
Temperature F	281.5	57.4	90.0	201.0	140.0	2662.1	100.0	260.0	412.2	382.3	100.0	86.7	100.0	100.0	110.0	320.0	282.7	110.3	247.0
Pressure psi	190.3	16.4	56.4	865.0	14.7	614.7	14.7	500.0	598.0	557.0	515.0	506.0	495.0	27.0	549.5	25.0	515.0	75.0	36.0
Mole Flow lbmol/hr	223,664	118,405	58,426	44,698	0	150,128	0	152,039	3,022	298,139	174,655	174,655	137,041	2,392	349	176	127,234	1,786	126,920
Mass Flow lb/hr	6,443,880	3,310,693	1,666,726	1,430,670	1,533,330	3,190,520	187,042	2,739,450	54,448	5,676,620	3,628,640	3,628,540	2,113,190	75,158	36,496	46,275	2,236,960	32,179	2,326,340
Volume Flow cuft/hr	9,357,680	40,018,600	8,210,000	478,637	0	3,235,280	0	68,218	1,262	4,624,100	2,008,380	2,047,430	1,671,350	527,961	8,118	142	46,169	315	172,139
Enthalpy MMBtu/hr	162.8	-172.0	5.0	34.1	-685.3	-2,626.0	-4.3	-18,250.5	-352.9	-21,220.5	-8,761.9	-8,761.2	-3,298.1	-89.3	-126.3	2.5	-15,169.8	-219.6	-15,489.4
Mole Frac																			
H2O	0.007	0.012	0.000	0.000	0.000	0.031	0.000	1.000	0.990	0.417	0.002	0.002	0.000	0.011	0.003	0.000	0.369	1.000	0.369
CO2	0.000	0.000	0.000	0.000	0.000	0.021	0.000	0.000	0.000	0.109	0.184	0.184	0.008	0.141	0.780	0.000	0.000	0.000	0.000
O2	0.207	0.015	0.005	0.950	0.000	0.000	0.000	0.000	0.000	0.000	0.000	0.000	0.000	0.000	0.000	0.000	0.000	0.000	0.000
N2	0.768	0.973	0.992	0.050	0.000	0.020	0.000	0.000	0.000	0.013	0.018	0.018	0.023	0.003	0.077	0.000	0.000	0.000	0.000
CH4	0.000	0.000	0.000	0.000	0.000	0.000	0.000	0.000	0.000	0.000	0.000	0.000	0.000	0.000	0.000	0.000	0.000	0.000	0.000
CO	0.000	0.000	0.000	0.000	0.000	0.350	0.000	0.000	0.000	0.231	0.395	0.395	0.480	0.150	0.000	0.000	0.000	0.000	0.000
COS	0.000	0.000	0.000	0.000	0.000	0.001	0.000	0.000	0.000	0.000	0.000	0.000	0.000	0.000	0.001	0.000	0.000	0.000	0.000
H2	0.000	0.000	0.000	0.000	0.000	0.264	0.000	0.000	0.000	0.222	0.393	0.393	0.489	0.092	0.089	0.000	0.000	0.000	0.000
H2S	0.000	0.000	0.000	0.000	0.000	0.009	0.000	0.000	0.000	0.005	0.009	0.009	0.000	0.602	0.049	0.000	0.000	0.000	0.000
HCL	0.000	0.000	0.000	0.000	0.000	0.001	0.000	0.000	0.000	0.001	0.000	0.000	0.000	0.000	0.000	0.000	0.001	0.000	0.001
S8	0.000	0.000	0.000	0.000	0.000	0.000	0.000	0.000	0.000	0.000	0.000	0.000	0.000	0.000	0.000	1.000	0.000	0.000	0.000
C2H4	0.000	0.000	0.000	0.000	0.000	0.000	0.000	0.000	0.000	0.000	0.000	0.000	0.000	0.000	0.000	0.000	0.000	0.000	0.000
C2H6	0.000	0.000	0.000	0.000	0.000	0.000	0.000	0.000	0.000	0.000	0.000	0.000	0.000	0.000	0.000	0.000	0.000	0.000	0.000
C3H6	0.000	0.000	0.000	0.000	0.000	0.000	0.000	0.000	0.000	0.000	0.000	0.000	0.000	0.000	0.000	0.000	0.000	0.000	0.000
C3H8	0.000	0.000	0.000	0.000	0.000	0.000	0.000	0.000	0.000	0.000	0.000	0.000	0.000	0.000	0.000	0.000	0.000	0.000	0.000
ISOBU-01	0.000	0.000	0.000	0.000	0.000	0.000	0.000	0.000	0.000	0.000	0.000	0.000	0.000	0.000	0.000	0.000	0.000	0.000	0.000
N-BUT-01	0.000	0.000	0.000	0.000	0.000	0.000	0.000	0.000	0.000	0.000	0.000	0.000	0.000	0.000	0.000	0.000	0.000	0.000	0.000
1-BUT-01	0.000	0.000	0.000	0.000	0.000	0.000	0.000	0.000	0.000	0.000	0.000	0.000	0.000	0.000	0.000	0.000	0.000	0.000	0.000
1-PEN-01	0.000	0.000	0.000	0.000	0.000	0.000	0.000	0.000	0.000	0.000	0.000	0.000	0.000	0.000	0.000	0.000	0.000	0.000	0.000
N-PEN-01	0.000	0.000	0.000	0.000	0.000	0.000	0.000	0.000	0.000	0.000	0.000	0.000	0.000	0.000	0.000	0.000	0.000	0.000	0.000
1-HEX-01	0.000	0.000	0.000	0.000	0.000	0.000	0.000	0.000	0.000	0.000	0.000	0.000	0.000	0.000	0.000	0.000	0.000	0.000	0.000
N-HEX-01	0.000	0.000	0.000	0.000	0.000	0.000	0.000	0.000	0.000	0.000	0.000	0.000	0.000	0.000	0.000	0.000	0.000	0.000	0.000
C7-C11	0.000	0.000	0.000	0.000	0.000	0.000	0.000	0.000	0.000	0.000	0.000	0.000	0.000	0.000	0.000	0.000	0.000	0.000	0.000
C12-C18	0.000	0.000	0.000	0.000	0.000	0.000	0.000	0.000	0.000	0.000	0.000	0.000	0.000	0.000	0.000	0.000	0.000	0.000	0.000
C16-C24	0.000	0.000	0.000	0.000	0.000	0.000	0.000	0.000	0.000	0.000	0.000	0.000	0.000	0.000	0.000	0.000	0.000	0.000	0.000
C25P_	0.000	0.000	0.000	0.000	0.000	0.000	0.000	0.000	0.000	0.000	0.000	0.000	0.000	0.000	0.000	0.000	0.000	0.000	0.000
ALCS	0.000	0.000	0.000	0.000	0.000	0.000	0.000	0.000	0.000	0.000	0.000	0.000	0.000	0.000	0.000	0.000	0.000	0.000	0.000
Solids Mass Flow lb/hr	6,443,880	3,310,693	1,666,726	1,430,670	1,533,330	3,190,520	187,042	2,739,450	54,448	5,676,620	3,628,640	3,628,540	2,113,190	75,158	36,496	46,275	2,236,960	32,178	2,326,340
Enthalpy MMBtu/hr	162.8	-172.0	5.0	34.1	-685.3	-2626.0	-4.3	-18250.5	-352.9	-21220.5	-8761.9	-8761.2	-3298.1	-89.3	-126.3	2.5	-15169.8	-219.6	-15489.4
Pressure psi	190.3	16.4	56.4	865.0	14.7	614.7	14.7	500.0	598.0	557.0	515.0	606.0	495.0	27.0	549.5	25.0	515.0	75.0	36.0
Temperature F					140		100												
Pressure psi	190.3	16.4	56.4	865.0	14.7	614.7	14.7	500.0	598.0	557.0	515.0	606.0	495.0	27.0	549.5	25.0	515.0	75.0	36.0
Mass Flow lb/hr	0	0	0	0	1,533,330	0	187,042	0	0	0	0	0	0	0	0	0	0	0	0
Enthalpy MMBtu/hr					-685.3		-4.3												

Table D-3. (Continued).

Stream #	20	21	22	23	24	25	26	27	28
Temperature F	500.0	214.1	406.4	100.0	100.0	100.0	463.4	323.5	100.0
Pressure psi	490.0	490.0	490.0	475.0	490.0	490.0	490.0	490.0	465.0
Mole Flow lbmol/hr	132,905	173,923	83,678	3,312	75,741	16,434	1,323	3,352	338
Mass Flow lb/hr	2,104,740	2,829,080	2,829,080	6,678	1,981,940	298,794	496,245	492,923	23,556
volume Flow cuft/hr	2,626,710	2,589,770	1,737,490	42,363	393,047	5,763	13,114	168,262	6,312
Enthalpy MMBtu/hr	-2,918.4	-3,302.0	-6,920.8	0.5	-4,663.0	-2,025.9	-396.5	-338.2	-26.3
Mole Frac									
H2O	0.000	0.000	0.175	0.000	0.000	0.995	0.000	0.000	0.000
CO2	0.009	0.015	0.262	0.000	0.349	0.000	0.000	0.000	0.000
O2	0.000	0.000	0.000	0.000	0.000	0.000	0.000	0.000	0.000
N2	0.023	0.089	0.165	0.000	0.234	0.000	0.000	0.000	0.000
CH4	0.000	0.008	0.014	0.000	0.017	0.000	0.000	0.062	0.310
CO	0.465	0.420	0.068	0.000	0.121	0.000	0.000	0.000	0.000
COS	0.000	0.000	0.000	0.000	0.000	0.000	0.000	0.000	0.000
H2	0.473	0.462	0.233	1.000	0.239	0.000	0.000	0.000	0.000
H2S	0.000	0.000	0.000	0.000	0.000	0.000	0.000	0.000	0.000
HCL	0.000	0.000	0.000	0.000	0.000	0.000	0.000	0.000	0.000
S8	0.000	0.000	0.000	0.000	0.000	0.000	0.000	0.000	0.000
C2H4	0.000	0.002	0.005	0.000	0.006	0.000	0.000	0.000	0.000
C2H6	0.000	0.001	0.002	0.000	0.002	0.000	0.000	0.046	0.230
C3H6	0.000	0.002	0.004	0.000	0.005	0.000	0.000	0.000	0.000
C3H8	0.000	0.001	0.001	0.000	0.002	0.000	0.000	0.046	0.230
ISOBU-01	0.000	0.000	0.000	0.000	0.000	0.000	0.000	0.023	0.116
N-BUT-01	0.000	0.001	0.001	0.000	0.001	0.000	0.000	0.023	0.115
1-BUT-01	0.000	0.002	0.004	0.000	0.004	0.000	0.000	0.000	0.000
1-PEN-01	0.000	0.000	0.001	0.000	0.000	0.000	0.000	0.000	0.000
N-PEN-01	0.000	0.000	0.000	0.000	0.000	0.000	0.000	0.046	0.000
1-HEX-01	0.000	0.000	0.001	0.000	0.000	0.000	0.000	0.000	0.000
N-HEX-01	0.000	0.000	0.000	0.000	0.000	0.000	0.000	0.046	0.000
C7-C11	0.000	0.000	0.002	0.000	0.000	0.000	0.000	0.255	0.000
C12-C18	0.000	0.000	0.001	0.000	0.000	0.000	0.000	0.344	0.000
C19-C24	0.000	0.000	0.001	0.000	0.000	0.000	0.766	0.111	0.000
C25PL	0.000	0.000	0.010	0.000	0.000	0.000	0.234	0.000	0.000
ALCS	0.000	0.000	0.001	0.000	0.000	0.005	0.000	0.000	0.000
Solids Mass Flow lb/hr	2,104,740	2,829,080	2,829,080	6,678	1,981,940	298,794	496,245	492,923	23,556
Enthalpy MMBtu/hr	-2918.4	-3302.0	-6920.8	0.5	-4663.0	-2025.9	-396.5	-338.2	-26.3
Pressure psi	490.0	490.0	490.0	475.0	490.0	490.0	490.0	490.0	465.0
Temperature F									
Pressure psi	490.0	490.0	490.0	475.0	490.0	490.0	490.0	490.0	465.0
Mass Flow lb/hr	0	0	0	0	0	0	0	0	0
Enthalpy MMBtu/hr									

Stream #	29	30	31	32	33	34	35	36	37
Temperature F	77.0	77.0	100.0	100.0	91.3	39.1	250.0	100.0	270.0
Pressure psi	14.7	14.7	490.0	490.0	465.0	20.0	15.6	490.0	14.7
Mole Flow lbmol/hr	1,582	1,834	1,888	356	8,098	53,430	36,387	460	82,400
Mass Flow lb/hr	160,236	371,239	34,741	15,116	154,762	2,294,850	1,133,590	62,106	2,313,790
volume Flow cuft/hr	3,684	8,013	24,073	10,474	101,620	14,172,000	16,303,700	1,426	43,889,400
Enthalpy MMBtu/hr	-146.6	-324.1	-31.1	-13.5	-145.6	-8,733.7	20.6	-49.5	-1,272.5
Mole Frac									
H2O	0.000	0.000	0.000	0.000	0.000	0.001	0.307	0.000	0.082
CO2	0.000	0.000	0.039	0.036	0.035	0.958	0.000	0.000	0.049
O2	0.000	0.000	0.000	0.000	0.000	0.000	0.207	0.000	0.026
N2	0.000	0.000	0.301	0.301	0.276	0.000	0.786	0.000	0.843
CH4	0.000	0.000	0.025	0.025	0.048	0.000	0.000	0.000	0.000
CO	0.000	0.000	0.179	0.176	0.184	0.028	0.000	0.000	0.000
COS	0.000	0.000	0.000	0.000	0.000	0.000	0.000	0.000	0.000
H2	0.000	0.000	0.426	0.426	0.391	0.013	0.000	0.000	0.000
H2S	0.000	0.000	0.000	0.000	0.000	0.000	0.000	0.000	0.000
HCL	0.000	0.000	0.000	0.000	0.000	0.000	0.000	0.000	0.000
S8	0.000	0.000	0.000	0.000	0.000	0.000	0.000	0.000	0.000
C2H4	0.000	0.000	0.009	0.009	0.008	0.000	0.000	0.000	0.000
C2H6	0.000	0.000	0.003	0.003	0.021	0.000	0.000	0.000	0.000
C3H6	0.000	0.000	0.008	0.008	0.007	0.000	0.000	0.000	0.000
C3H8	0.000	0.000	0.002	0.002	0.021	0.000	0.000	0.000	0.000
ISOBU-01	0.000	0.000	0.000	0.000	0.009	0.000	0.000	0.000	0.000
N-BUT-01	0.000	0.000	0.002	0.002	0.011	0.000	0.000	0.000	0.000
1-BUT-01	0.000	0.000	0.007	0.007	0.006	0.000	0.000	0.000	0.000
1-PEN-01	0.046	0.000	0.000	0.000	0.000	0.000	0.000	0.146	0.000
N-PEN-01	0.106	0.000	0.000	0.000	0.000	0.000	0.000	0.016	0.000
1-HEX-01	0.039	0.000	0.000	0.000	0.000	0.000	0.000	0.126	0.000
N-HEX-01	0.105	0.000	0.000	0.000	0.000	0.000	0.000	0.014	0.000
C7-C11	0.706	0.000	0.000	0.000	0.000	0.000	0.000	0.473	0.000
C12-C18	0.000	0.772	0.000	0.000	0.000	0.000	0.000	0.228	0.000
C19-C24	0.000	0.228	0.000	0.000	0.000	0.000	0.000	0.000	0.000
C25PL	0.000	0.000	0.000	0.000	0.000	0.000	0.000	0.000	0.000
ALCS	0.000	0.000	0.000	0.000	0.000	0.000	0.000	0.000	0.000
Solids Mass Flow lb/hr	160,236	371,239	34,741	15,116	154,762	2,294,850	1,133,590	62,106	2,313,790
Enthalpy MMBtu/hr	-146.6	-324.1	-31.1	-13.5	-145.6	-8733.7	20.6	-49.5	-1272.5
Pressure psi	14.7	14.7		490.0	465.0	20.0	15.6	490.0	14.7
Temperature F									
Pressure psi	14.7	14.7		490.0	465.0	20.0	15.6	490.0	14.7
Mass Flow lb/hr	0	0	0	0	0	0	0	0	0
Enthalpy MMBtu/hr									

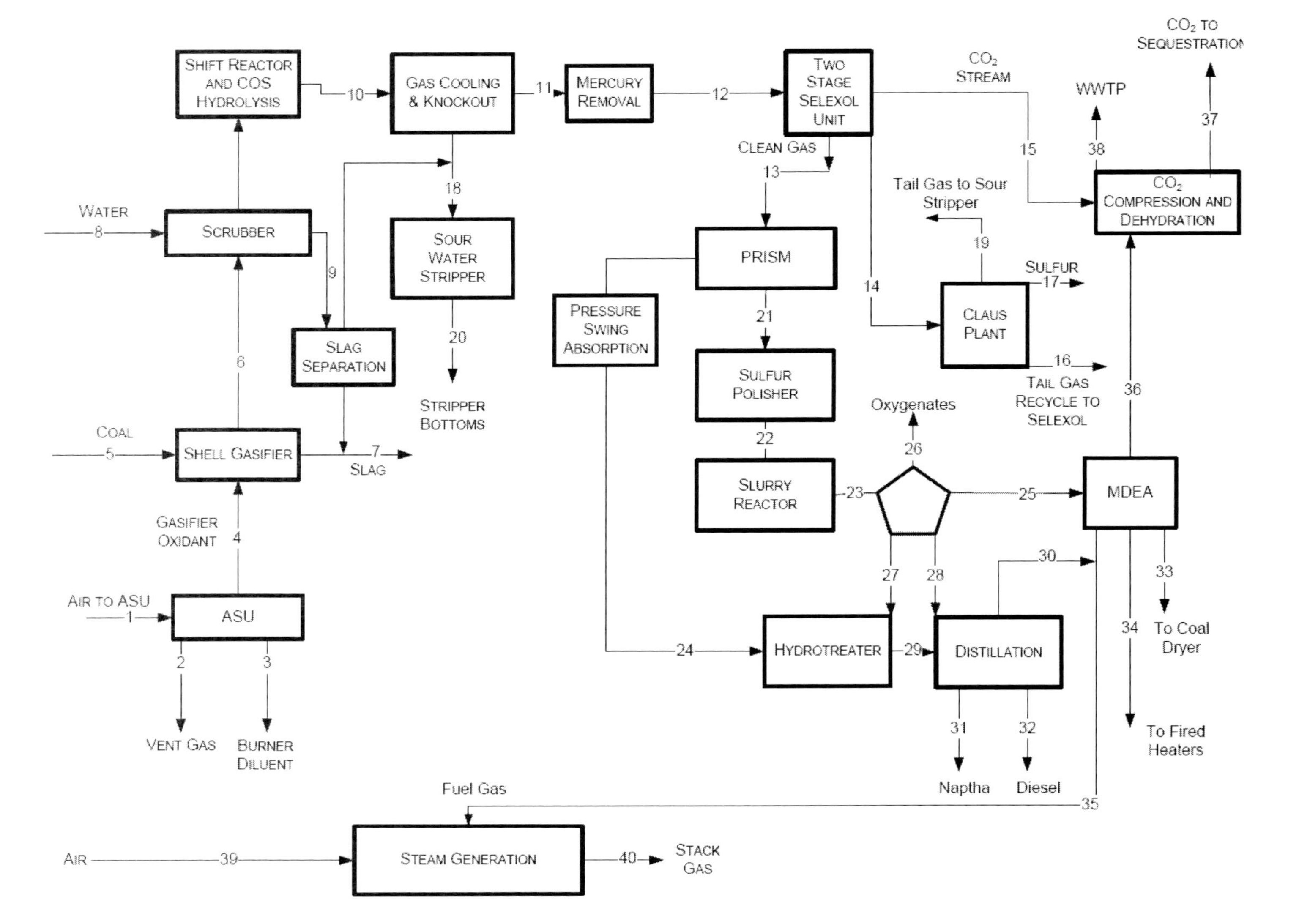

Figure D-2. CTL with CCS (Case 2).

Table D-4. CTL with CCS (Case 2)

Visio Stream #	1	2	3	4	5	6	8	7	9	10	11	12	13	14	15	16	17	18	19	20
Temperature F	261.6	57.4	93.0	201.0	140.0	2682.1	288.0	100.0	412.3	382.7	100.8	99.7	100.0	100.0	176.8	110.0	320.0	291.9	110.3	347.0
Pressure psi	160.3	16.4	53.1	865.0	14.7	614.7	308.0	14.7	596.0	567.0	516.8	505.0	495.0	27.0	2214.7	549.5	25.0	515.0	75.0	35.0
Mole Flow lbmol/hr	218,740	116,852	58,045	43,949	0	146,627	143,522	0	2,961	292,197	170,338	170,338	135,950	2,339	26,245	925	172	124,016	1,745	125,867
Mass Flow lb/hr	6,293,860	3,233,610	1,627,907	1,397,580	1,853,910	3,116,260	2,675,680	182,693	53,179	5,738,760	3,548,380	3,548,380	2,057,710	73,495	1,155,030	35,690	44,221	2,237,930	31,444	2,266,620
Volume Flow cuft/hr	8,159,330	39,082,100	6,070,000	465,939	0	6,043,260	54,910	0	1,223	4,418,450	1,984,240	2,002,520	1,636,070	516,125	44,201	8,914	138	46,925	801	166,061
Enthalpy MMBtu/hr	158.0	-105.0	5.0	33.3	-962.4	-2,585.7	-17,325.6	-4.2	-344.7	-20,734.0	-8,568.9	-8,568.9	-3,210.7	-87.6	-4,489.1	-123.5	2.4	-14,756.3	-214.6	-15,086.2
Mole Frac																				
H_2O	0.007	0.012	0.000	0.000	0.000	0.031	1.000	0.000	0.999	0.416	0.002	0.002	0.000	0.011	0.000	0.003	0.000	0.998	1.000	0.999
CO_2	0.000	0.000	0.000	0.000	0.000	0.021	0.000	0.000	0.000	0.109	0.185	0.185	0.008	0.142	1.000	0.781	0.000	0.000	0.000	0.000
O_2	0.207	0.018	0.008	0.950	0.000	0.000	0.000	0.000	0.000	0.000	0.000	0.000	0.000	0.000	0.000	0.000	0.000	0.000	0.000	0.000
N_2	0.786	0.970	0.992	0.050	0.000	0.020	0.000	0.000	0.000	0.010	0.018	0.018	0.023	0.003	0.000	0.077	0.000	0.000	0.000	0.000
CH_4	0.000	0.000	0.000	0.000	0.000	0.000	0.000	0.000	0.000	0.000	0.000	0.000	0.000	0.000	0.000	0.000	0.000	0.000	0.000	0.000
CO	0.000	0.000	0.000	0.000	0.000	0.653	0.000	0.000	0.000	0.230	0.393	0.393	0.479	0.150	0.000	0.000	0.000	0.000	0.000	0.000
COS	0.000	0.000	0.000	0.000	0.000	0.001	0.000	0.000	0.000	0.000	0.000	0.000	0.000	0.000	0.000	0.001	0.000	0.000	0.000	0.000
H_2	0.000	0.000	0.000	0.000	0.000	0.264	0.000	0.000	0.000	0.230	0.394	0.394	0.491	0.093	0.000	0.069	0.000	0.000	0.000	0.000
H_2S	0.000	0.000	0.000	0.000	0.000	0.009	0.000	0.000	0.000	0.005	0.006	0.006	0.000	0.602	0.000	0.049	0.000	0.000	0.000	0.000
HCL	0.000	0.000	0.000	0.000	0.000	0.001	0.000	0.000	0.000	0.001	0.000	0.000	0.000	0.000	0.000	0.000	0.000	0.001	0.000	0.001
S8	0.000	0.000	0.000	0.000	0.000	0.000	0.000	0.000	0.000	0.000	0.000	0.000	0.000	0.000	0.000	0.000	1.000	0.000	0.000	0.000
C_2H_4	0.000	0.000	0.000	0.000	0.000	0.000	0.000	0.000	0.000	0.000	0.000	0.000	0.000	0.000	0.000	0.000	0.000	0.000	0.000	0.000
C_2H_6	0.000	0.000	0.000	0.000	0.000	0.000	0.000	0.000	0.000	0.000	0.000	0.000	0.000	0.000	0.000	0.000	0.000	0.000	0.000	0.000
C_3H_6	0.000	0.000	0.000	0.000	0.000	0.000	0.000	0.000	0.000	0.000	0.000	0.000	0.000	0.000	0.000	0.000	0.000	0.000	0.000	0.000
C_3H_8	0.000	0.000	0.000	0.000	0.000	0.000	0.000	0.000	0.000	0.000	0.000	0.000	0.000	0.000	0.000	0.000	0.000	0.000	0.000	0.000
ISOBU-01	0.000	0.000	0.000	0.000	0.000	0.000	0.000	0.000	0.000	0.000	0.000	0.000	0.000	0.000	0.000	0.000	0.000	0.000	0.000	0.000
N-BUT-01	0.000	0.000	0.000	0.000	0.000	0.000	0.000	0.000	0.000	0.000	0.000	0.000	0.000	0.000	0.000	0.000	0.000	0.000	0.000	0.000
1-BUT-01	0.000	0.000	0.000	0.000	0.000	0.000	0.000	0.000	0.000	0.000	0.000	0.000	0.000	0.000	0.000	0.000	0.000	0.000	0.000	0.000
1-PEN-01	0.000	0.000	0.000	0.000	0.000	0.000	0.000	0.000	0.000	0.000	0.000	0.000	0.000	0.000	0.000	0.000	0.000	0.000	0.000	0.000
N-PEN-01	0.000	0.000	0.000	0.000	0.000	0.000	0.000	0.000	0.000	0.000	0.000	0.000	0.000	0.000	0.000	0.000	0.000	0.000	0.000	0.000
1-HEX-01	0.000	0.000	0.000	0.000	0.000	0.000	0.000	0.000	0.000	0.000	0.000	0.000	0.000	0.000	0.000	0.000	0.000	0.000	0.000	0.000
N-HEX-01	0.000	0.000	0.000	0.000	0.000	0.000	0.000	0.000	0.000	0.000	0.000	0.000	0.000	0.000	0.000	0.000	0.000	0.000	0.000	0.000
C7-C11	0.000	0.000	0.000	0.000	0.000	0.000	0.000	0.000	0.000	0.000	0.000	0.000	0.000	0.000	0.000	0.000	0.000	0.000	0.000	0.000
C12-C18	0.000	0.000	0.000	0.000	0.000	0.000	0.000	0.000	0.000	0.000	0.000	0.000	0.000	0.000	0.000	0.000	0.000	0.000	0.000	0.000
C19-C24	0.000	0.000	0.000	0.000	0.000	0.000	0.000	0.000	0.000	0.000	0.000	0.000	0.000	0.000	0.000	0.000	0.000	0.000	0.000	0.000
C25PL	0.000	0.000	0.000	0.000	0.000	0.000	0.000	0.000	0.000	0.000	0.000	0.000	0.000	0.000	0.000	0.000	0.000	0.000	0.000	0.000
ALCS	0.000	0.000	0.000	0.000	0.000	0.000	0.000	0.000	0.000	0.000	0.000	0.000	0.000	0.000	0.000	0.000	0.000	0.000	0.000	0.000
Solids Mass Flow lb/hr	6,293,860	3,233,610	1,650,000	1,397,580	1,853,910	3,116,260	2,675,680	182,693	53,179	5,738,760	3,548,380	3,548,380	2,057,710	73,495	1,155,030	35,690	44,221	2,237,930	31,444	2,266,620
Enthalpy MMBtu/hr	158.0	-105.0	4.6	33.3	-962.4	-2585.7	-17325.6	-4.2	-344.7	-20734.0	-8568.9	-8569.9	-3210.7	-87.6	-4489.1	-123.5	2.4	-14756.3	-214.6	-15098.2
Pressure psi	160.3	16.4	53.4	865.0	14.7	614.7	300.0	14.7	596.0	567.0	515.8	505.0	495.0	27.0	2214.7	549.5	25.0	515.0	75.0	35.0
Temperature F					140.0			100.0												
Pressure psi	160.3	16.4	53.4	865.0	14.7	614.7	300.0	14.7	596.0	567.0	515.8	505.0	495.0	27.0	2214.7	549.5	25.0	515.0	75.0	35.0
Mass Flow lb/hr	0	0	0	0	1,853,910	0	0	182,693	0	0	0	0	0	0	0	0	0	0	0	0
Enthalpy MMBtu/hr					-962.4			-4.2												

Stream #	21	22	23	24	25	26	27	28	29	30
Temperature F	500.0	211.1	405.7	100.0	100.0	100.0	455.4	100.0	323.5	100.0
Pressure psi	490.0	460.0	490.0	475.0	490.0	490.0	490.0	490.0	460.0	485.0
Mole Flow lbmol/hr	129,714	173,412	93,967	3,312	75,730	18,717	1,023	460	3,351	665
Mass Flow lb/hr	2,049,250	2,325,190	2,626,180	6,678	1,973,540	333,877	435,715	62,239	492,394	23,533
Volume Flow cuft/hr	2,763,810	2,570,300	1,742,870	42,353	983,091	5,850	13,114	1,426	158,322	5,308
Enthalpy MMBtu/hr	-2,340.1	-3,767.5	-3,793.0	0.5	-4,597.2	-2,060.7	-386.5	-49.5	-366.2	-28.2
Mole Frac										
H2O	0.000	0.000	0.177	0.000	0.000	0.995	0.000	0.000	0.000	0.000
CO2	0.003	0.015	0.278	0.000	0.346	0.000	0.000	0.000	0.000	0.000
O2	0.000	0.000	0.000	0.000	0.000	0.000	0.000	0.000	0.000	0.000
N2	0.023	0.096	0.165	0.000	0.204	0.000	0.000	0.000	0.000	0.000
CH4	0.000	0.007	0.015	0.000	0.016	0.000	0.000	0.000	0.082	0.313
CO	0.493	0.418	0.085	0.000	0.115	0.000	0.000	0.000	0.000	0.000
COS	0.000	0.000	0.000	0.000	0.000	0.000	0.000	0.000	0.000	0.000
H2	0.475	0.461	0.234	1.000	0.261	0.000	0.000	0.000	0.000	0.000
H2S	0.000	0.000	0.000	0.000	0.000	0.000	0.000	0.000	0.000	0.000
HCL	0.000	0.000	0.000	0.000	0.000	0.000	0.000	0.000	0.000	0.000
S8	0.000	0.000	0.000	0.000	0.000	0.000	0.000	0.000	0.000	0.000
C2H4	0.000	0.002	0.005	0.000	0.007	0.000	0.000	0.000	0.000	0.000
C2H6	0.000	0.001	0.002	0.000	0.002	0.000	0.000	0.000	0.046	0.233
C3H6	0.000	0.002	0.005	0.000	0.006	0.000	0.000	0.000	0.000	0.000
C3H8	0.000	0.001	0.001	0.000	0.002	0.000	0.000	0.000	0.046	0.233
ISOBU-01	0.000	0.000	0.000	0.000	0.000	0.000	0.000	0.000	0.023	0.115
N-BUT-01	0.000	0.001	0.001	0.000	0.002	0.000	0.000	0.000	0.023	0.115
1-BUT-01	0.000	0.002	0.004	0.000	0.005	0.000	0.000	0.000	0.000	0.000
1-PEN-01	0.000	0.000	0.001	0.000	0.000	0.000	0.000	0.146	0.000	0.000
N-PEN-01	0.000	0.000	0.000	0.000	0.000	0.000	0.000	0.016	0.046	0.000
1-HEX-01	0.000	0.000	0.001	0.000	0.000	0.000	0.000	0.124	0.000	0.000
N-HEX-01	0.000	0.000	0.000	0.000	0.000	0.000	0.000	0.014	0.046	0.000
C7-C11	0.000	0.000	0.002	0.000	0.000	0.000	0.000	0.473	0.255	0.000
C12-C18	0.000	0.000	0.001	0.000	0.000	0.000	0.000	0.226	0.344	0.000
C19-C24	0.000	0.000	0.001	0.000	0.000	0.000	0.066	0.000	0.111	0.000
C25PL	0.000	0.000	0.010	0.000	0.000	0.000	0.934	0.000	0.000	0.000
ALC3	0.000	0.000	0.001	0.000	0.000	0.005	0.000	0.000	0.000	0.000
Solids Mass Flow lb/hr	2,049,250	2,325,190	2,626,180	6,678	1,973,540	333,877	435,715	62,239	492,394	23,533
Enthalpy MMBtu/hr	-2340.1	-3767.5	-3793.0	0.5	-4597.2	-2060.7	-386.5	-49.5	-366.2	-28.2
Pressure psi	490.0	460.0	490.0	475.0	490.0	490.0	490.0	490.0	460.0	485.0
Temperature F										
Pressure psi	490.0	460.0	490.0	475.0	490.0	490.0	490.0	490.0	460.0	485.0
Mass Flow lb/hr	0	0	0	0	0	0	0	0	0	0
Enthalpy MMBtu/hr										

Stream #	31	32	33	34	35	36	37	38	39	40
Temperature F	77.0	77.0	100.0	130.0	91.4	105.0	141.4	82.9	250.0	270.0
Pressure psi	14.7	14.7	490.0	490.0	485.0	2214.7	2214.7	20.0	16.5	14.7
Mole Flow lbmol/hr	1,530	1,336	1,870	827	8,284	24,152	50,427	46	41,692	83,807
Mass Flow lb/hr	180,083	370,340	32,977	14,693	164,396	1,094,170	2,219,200	823	1,198,030	2,354,030
Volume Flow cuft/hr	3,680	6,017	22,865	10,113	101,413	22,830	83,282	16	20,402,000	44,639,300
Enthalpy MMBtu/hr	-143.3	-324.3	-26.1	-12.9	-143.5	-4,160.0	-8,676.2	-5.6	21.9	-1,298.9
Mole Frac										
H2O	0.000	0.000	0.000	0.000	0.000	0.000	0.000	1.000	0.057	0.082
CO2	0.000	0.000	0.039	0.039	0.035	1.000	1.000	0.000	0.000	0.049
O2	0.000	0.000	0.000	0.000	0.000	0.000	0.000	0.000	0.207	0.029
N2	0.000	0.000	0.300	0.300	0.276	0.000	0.000	0.000	0.735	0.840
CH4	0.000	0.000	0.027	0.027	0.051	0.000	0.000	0.000	0.000	0.000
CO	0.000	0.000	0.174	0.174	0.160	0.000	0.000	0.000	0.000	0.000
COS	0.000	0.000	0.000	0.000	0.000	0.000	0.000	0.000	0.000	0.000
H2	0.000	0.000	0.427	0.427	0.392	0.000	0.000	0.000	0.000	0.000
H2S	0.000	0.000	0.000	0.000	0.000	0.000	0.000	0.000	0.000	0.000
HCL	0.000	0.000	0.000	0.000	0.000	0.000	0.000	0.000	0.000	0.000
S8	0.000	0.000	0.000	0.000	0.000	0.000	0.000	0.000	0.000	0.000
C2H4	0.000	0.000	0.010	0.013	0.009	0.000	0.000	0.000	0.000	0.000
C2H6	0.000	0.000	0.003	0.003	0.022	0.000	0.000	0.000	0.000	0.000
C3H6	0.000	0.000	0.008	0.009	0.006	0.000	0.000	0.000	0.000	0.000
C3H8	0.000	0.000	0.003	0.003	0.021	0.000	0.000	0.000	0.000	0.000
ISOBU-01	0.000	0.000	0.000	0.000	0.009	0.000	0.000	0.000	0.000	0.000
N-BUT-01	0.000	0.000	0.002	0.002	0.012	0.000	0.000	0.000	0.000	0.000
1-BUT-01	0.000	0.000	0.007	0.007	0.007	0.000	0.000	0.000	0.000	0.000
1-PEN-01	0.045	0.000	0.000	0.000	0.000	0.000	0.000	0.000	0.000	0.000
N-PEN-01	0.105	0.000	0.000	0.000	0.000	0.000	0.000	0.000	0.000	0.000
1-HEX-01	0.039	0.000	0.000	0.000	0.000	0.000	0.000	0.000	0.000	0.000
N-HEX-01	0.105	0.000	0.000	0.000	0.000	0.000	0.000	0.000	0.000	0.000
C7-C11	0.706	0.000	0.000	0.000	0.000	0.000	0.000	0.000	0.000	0.000
C12-C18	0.000	0.772	0.000	0.000	0.000	0.000	0.000	0.000	0.000	0.000
C19-C24	0.000	0.228	0.000	0.000	0.000	0.000	0.000	0.000	0.000	0.000
C25PL	0.000	0.000	0.000	0.000	0.000	0.000	0.000	0.000	0.000	0.000
ALC3	0.000	0.000	0.000	0.000	0.000	0.000	0.000	0.000	0.000	0.000
Solids Mass Flow lb/hr	180,083	370,340	32,977	14,693	164,396	1,050,030	2,219,200	823	1,198,030	2,354,030
Enthalpy MMBtu/hr	-143.3	-324.3	-26.1	-12.9	-143.5	-4160.1	-8676.2	-5.6	21.9	-1298.9
Pressure psi	14.7	14.7		490.0	485.0	2214.7	2214.7	20.0	16.5	14.7
Temperature F										
Pressure psi	14.7	14.7		490.0	485.0	2214.7	2214.7	20.0	16.5	14.7
Mass Flow lb/hr	0	0	0	0	0	0	0	0	0	0
Enthalpy MMBtu/hr										

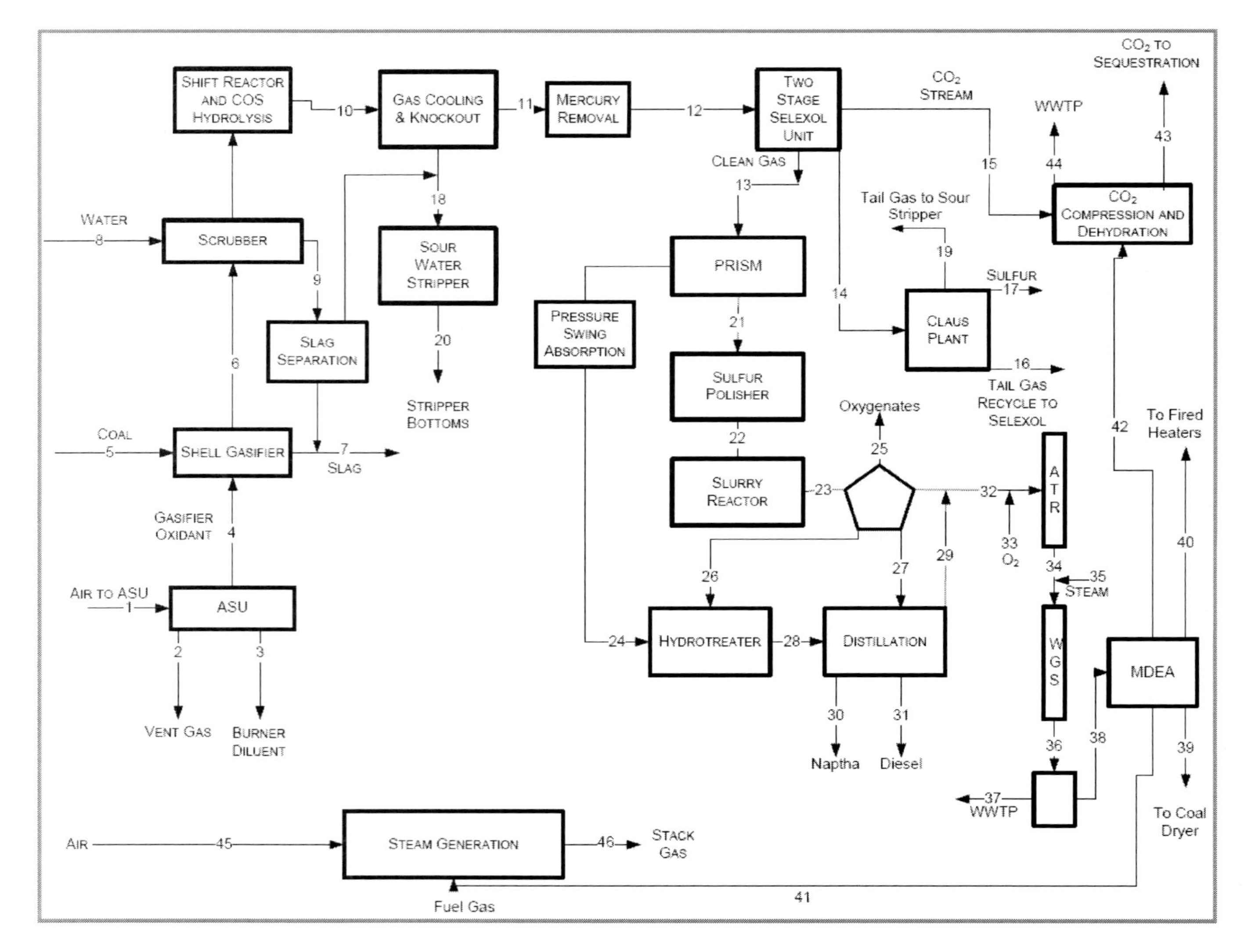

Figure D-3. CTL with CCS+ATR (Case 3).

Table D-5. CTL with CCS+ATR (Case 3)

Stream #	1	2	3	4	5	6	7	8	9	10	11	12	13	14	15	16
Temperature F	281.5	57.4	90.0	201.0	140.0	2631.9	100.0	260.0	413.5	377.1	100.0	99.7	100.0	100.0	175.6	110.0
Pressure psi	190.3	16.4	56.4	635.0	14.7	614.7	14.7	614.7	612.7	571.7	529.7	519.7	509.7	27.0	2214.7	549.5
Mole Flow lbmol/hr	270,644	145,631	69,282	46,723	0	158,891	0	160,035	3,193	315,787	193,381	193,381	160,932	2,746	42,904	1,173
Mass Flow lb/hr	7,787,300	4,072,030	1,943,043	1,485,930	1,795,910	3,352,180	198,373	2,883,990	57,472	6,178,690	3,969,400	3,969,400	2,384,180	87,115	1,883,180	45,120
Volume Flow cuft/hr	11,345,100	49,221,200	7,240,000	495,289	0	6,715,410	0	59,133	1,323	4,800,780	2,156,280	2,197,130	1,907,810	606,178	72,257	11,188
Enthalpy MMBtu/hr	196.7	-207.7	5.4	35.4	-353.4	-2,618.3	-4.6	-19,213.4	-372.4	-22,376.5	-10,119.9	-10,119.9	-3,815.0	-122.0	-7,338.5	-162.0
Mole Frac																
H2O	0.007	0.012	0.000	0.000	0.000	0.025	0.000	1.000	0.999	0.388	0.002	0.002	0.000	0.011	0.000	0.003
CO2	0.000	0.000	0.000	0.000	0.000	0.017	0.000	0.000	0.000	0.130	0.212	0.212	0.011	0.188	1.000	0.811
O2	0.207	0.013	0.008	0.950	0.000	0.000	0.000	0.000	0.000	0.000	0.000	0.000	0.000	0.000	0.000	0.000
N2	0.786	0.973	0.992	0.050	0.000	0.020	0.000	0.000	0.000	0.010	0.017	0.017	0.038	0.004	0.000	0.072
CH4	0.000	0.000	0.000	0.000	0.000	0.000	0.000	0.000	0.000	0.000	0.000	0.000	0.000	0.000	0.000	0.001
CO	0.000	0.000	0.000	0.000	0.000	0.657	0.000	0.000	0.000	0.210	0.343	0.343	0.437	0.140	0.000	0.000
COS	0.000	0.000	0.000	0.000	0.000	0.001	0.000	0.000	0.000	0.000	0.000	0.000	0.000	0.000	0.000	0.001
H2	0.000	0.000	0.000	0.000	0.000	0.270	0.000	0.000	0.000	0.257	0.419	0.419	0.514	0.099	0.000	0.073
H2S	0.000	0.000	0.000	0.000	0.000	0.009	0.000	0.000	0.000	0.005	0.008	0.008	0.000	0.556	0.000	0.041
HCL	0.000	0.000	0.000	0.000	0.000	0.001	0.000	0.000	0.000	0.001	0.000	0.000	0.000	0.000	0.000	0.000
S8	0.000	0.000	0.000	0.000	0.000	0.000	0.000	0.000	0.000	0.000	0.000	0.000	0.000	0.000	0.000	0.000
C2H4	0.000	0.000	0.000	0.000	0.000	0.000	0.000	0.000	0.000	0.000	0.000	0.000	0.000	0.000	0.000	0.000
C2H6	0.000	0.000	0.000	0.000	0.000	0.000	0.000	0.000	0.000	0.000	0.000	0.000	0.000	0.000	0.000	0.000
C3H6	0.000	0.000	0.000	0.000	0.000	0.000	0.000	0.000	0.000	0.000	0.000	0.000	0.000	0.000	0.000	0.000
C3H8	0.000	0.000	0.000	0.000	0.000	0.000	0.000	0.000	0.000	0.000	0.000	0.000	0.000	0.000	0.000	0.000
ISOBU-01	0.000	0.000	0.000	0.000	0.000	0.000	0.000	0.000	0.000	0.000	0.000	0.000	0.000	0.000	0.000	0.000
N-BUT-01	0.000	0.000	0.000	0.000	0.000	0.000	0.000	0.000	0.000	0.000	0.000	0.000	0.000	0.000	0.000	0.000
1-BUT-01	0.000	0.000	0.000	0.000	0.000	0.000	0.000	0.000	0.000	0.000	0.000	0.000	0.000	0.000	0.000	0.000
1-PEN-01	0.000	0.000	0.000	0.000	0.000	0.000	0.000	0.000	0.000	0.000	0.000	0.000	0.000	0.000	0.000	0.000
N-PEN-01	0.000	0.000	0.000	0.000	0.000	0.000	0.000	0.000	0.000	0.000	0.000	0.000	0.000	0.000	0.000	0.000
1-HEX-01	0.000	0.000	0.000	0.000	0.000	0.000	0.000	0.000	0.000	0.000	0.000	0.000	0.000	0.000	0.000	0.000
N-HEX-01	0.000	0.000	0.000	0.000	0.000	0.000	0.000	0.000	0.000	0.000	0.000	0.000	0.000	0.000	0.000	0.000
C7-C11	0.000	0.000	0.000	0.000	0.000	0.000	0.000	0.000	0.000	0.000	0.000	0.000	0.000	0.000	0.000	0.000
C12-C18	0.000	0.000	0.000	0.000	0.000	0.000	0.000	0.000	0.000	0.000	0.000	0.000	0.000	0.000	0.000	0.000
C19-C24	0.000	0.000	0.000	0.000	0.000	0.000	0.000	0.000	0.000	0.000	0.000	0.000	0.000	0.000	0.000	0.000
C25PL	0.000	0.000	0.000	0.000	0.000	0.000	0.000	0.000	0.000	0.000	0.000	0.000	0.000	0.000	0.000	0.000
ALCS	0.000	0.000	0.000	0.000	0.000	0.000	0.000	0.000	0.000	0.000	0.000	0.000	0.000	0.000	0.000	0.000
Solids Mass Flow lb/hr	7,787,300	4,072,030	1,940,000	1,485,930	1,795,910	3,352,180	198,373	2,883,990	57,472	6,178,690	3,969,400	3,969,400	2,384,180	87,115	1,883,180	45,120
Enthalpy MMBtu/hr	196.7	-207.7	5.4	35.4	-353.4	-2618.3	-4.6	-19213.4	-372.4	-22376.5	-10119.9	-10119.9	-3815.0	-122.0	-7338.5	-162.0
Pressure psi	190.3	16.4	56.4	635.0	14.7	614.7	14.7	614.7	612.7	571.7	529.7	519.7	509.7	27.0	2214.7	549.5
Temperature F					140.0		100.0									
Pressure psi	190.3	16.4	56.4	635.0	14.7	614.7	14.7	614.7	612.7	571.7	529.7	519.7	509.7	27.0	2214.7	549.5
Mass Flow lb/hr	0	0	0	0	1,795,910	0	198,373	0	0	0	0	0	0	0	0	0
Enthalpy MMBtu/hr					-353.4		-4.6									

Table D-5. (Continued).

Stream #	17	18	19	20	21	22	23	24	25	26	27	28	29	30	31
Temperature F	320.0	289.2	110.3	247.0	500.0	242.5	398.1	100.0	100.0	452.3	100.0	321.6	100.0	77.0	77.0
Pressure psi	25.0	529.7	75.0	35.0	504.7	504.7	504.7	489.7	504.7	504.7	504.7	504.7	479.7	14.7	14.7
Mole Flow lbmol/hr	187	125,277	1,930	127,103	156,795	160,049	80,971	3,312	16,610	1,023	479	3,349	664	1,528	1,636
Mass Flow lb/hr	48,021	2,251,020	34,769	2,232,720	2,375,580	2,437,890	2,437,890	6,673	301,956	434,896	61,934	491,575	23,491	159,795	370,222
Volume Flow cuft/hr	150	47,335	665	130,666	3,238,400	2,415,560	1,431,560	41,097	5,813	13,103	1,424	159,189	4,976	3,674	8,023
Enthalpy MMBtu/hr	2.6	-14,944.6	-237.3	-15,269.1	-3,166.7	-3,551.2	-6,620.3	0.5	-2,047.6	-396.3	-49.5	-366.4	-28.2	-146.1	-324.5
Mole Frac															
H2O	0.000	0.998	1.000	0.999	0.000	0.000	0.204	0.000	0.995	0.000	0.000	0.000	0.000	0.000	0.000
CO2	0.000	0.000	0.000	0.000	0.311	0.011	0.311	0.000	0.000	0.000	0.000	0.000	0.000	0.000	0.000
O2	0.000	0.000	0.000	0.000	0.000	0.000	0.000	0.000	0.000	0.000	0.000	0.000	0.000	0.000	0.000
N2	0.000	0.000	0.000	0.000	0.039	0.038	0.076	0.000	0.000	0.000	0.000	0.000	0.000	0.000	0.000
CH4	0.000	0.000	0.000	0.000	0.000	0.000	0.003	0.000	0.000	0.000	0.000	0.061	0.310	0.000	0.000
CO	0.000	0.000	0.000	0.000	0.449	0.453	0.108	0.000	0.000	0.000	0.000	0.000	0.000	0.000	0.000
COS	0.000	0.000	0.000	0.000	0.000	0.000	0.000	0.000	0.000	0.000	0.000	0.000	0.000	0.000	0.000
H2	0.000	0.000	0.000	0.000	0.501	0.498	0.274	1.000	0.000	0.000	0.000	0.000	0.000	0.000	0.000
H2S	0.000	0.000	0.000	0.000	0.000	0.000	0.000	0.000	0.000	0.000	0.000	0.000	0.000	0.000	0.000
HCL	0.000	0.001	0.000	0.001	0.000	0.000	0.000	0.000	0.000	0.000	0.000	0.000	0.000	0.000	0.000
S8	1.000	0.000	0.000	0.000	0.000	0.000	0.000	0.000	0.000	0.000	0.000	0.000	0.000	0.000	0.000
C2H4	0.000	0.000	0.000	0.000	0.000	0.000	0.001	0.000	0.000	0.000	0.000	0.000	0.000	0.000	0.000
C2H6	0.000	0.000	0.000	0.000	0.000	0.000	0.000	0.000	0.000	0.000	0.000	0.046	0.230	0.000	0.000
C3H6	0.000	0.000	0.000	0.000	0.000	0.000	0.001	0.000	0.000	0.000	0.000	0.000	0.000	0.000	0.000
C3H8	0.000	0.000	0.000	0.000	0.000	0.000	0.000	0.000	0.000	0.000	0.000	0.046	0.230	0.000	0.000
ISOBU-01	0.000	0.000	0.000	0.000	0.000	0.000	0.000	0.000	0.000	0.000	0.000	0.023	0.115	0.000	0.000
N-BUT-01	0.000	0.000	0.000	0.000	0.000	0.000	0.000	0.000	0.000	0.000	0.000	0.023	0.115	0.000	0.000
1-BUT-01	0.000	0.000	0.000	0.000	0.000	0.000	0.001	0.000	0.000	0.000	0.000	0.000	0.000	0.000	0.000
1-PEN-01	0.000	0.000	0.000	0.000	0.000	0.000	0.001	0.000	0.000	0.000	0.145	0.000	0.000	0.045	0.000
N-PEN-01	0.000	0.000	0.000	0.000	0.000	0.000	0.000	0.000	0.000	0.000	0.016	0.046	0.000	0.105	0.000
1-HEX-01	0.000	0.000	0.000	0.000	0.000	0.000	0.001	0.000	0.000	0.000	0.124	0.000	0.000	0.039	0.000
N-HEX-01	0.000	0.000	0.000	0.000	0.000	0.000	0.000	0.000	0.000	0.000	0.013	0.046	0.000	0.105	0.000
C7-C11	0.000	0.000	0.000	0.000	0.000	0.000	0.003	0.000	0.000	0.000	0.473	0.254	0.000	0.706	0.000
C12-C18	0.000	0.000	0.000	0.000	0.000	0.000	0.001	0.000	0.000	0.000	0.229	0.345	0.000	0.000	0.772
C19-C24	0.000	0.000	0.000	0.000	0.000	0.000	0.001	0.000	0.000	0.066	0.000	0.111	0.000	0.000	0.228
C25PL	0.000	0.000	0.000	0.000	0.000	0.000	0.012	0.000	0.000	0.934	0.000	0.000	0.000	0.000	0.000
ALCS	0.000	0.000	0.000	0.000	0.000	0.000	0.001	0.000	0.005	0.000	0.000	0.000	0.000	0.000	0.000
Solids Mass Flow lb/hr	48,021	2,251,020	34,769	2,232,720	2,375,580	2,437,890	2,437,890	6,673	301,956	434,896	61,934	491,575	23,491	159,795	370,222
Enthalpy MMBtu/hr	2.6	-14944.6	-237.3	-15269.1	-3166.7	-3551.2	-6620.3	0.5	-2047.6	-396.3	-49.5	-366.4	-28.2	-146.1	-324.5
Pressure psi	25.0	529.7	75.0	35.0	504.7	504.7	504.7	489.7	504.7	504.7	504.7	504.7	479.7	14.7	14.7
Temperature F															
Pressure psi	25.0	529.7	75.0	35.0	504.7	504.7	504.7	489.7	504.7	504.7	504.7	504.7	479.7	14.7	14.7
Mass Flow lb/hr	0	0	0	0	0	0	0	0	0	0	0	0	0	0	0
Enthalpy MMBtu/hr															

Stream #	32	33	34	35	36	37	38	39	40	41	42	43	44	45	46
Temperature F	97.9	177.8	671.1	490.0	309.6	105.0	105.0	105.0	105.0	105.0	105.0	153.6	63.6	250.0	270.0
Pressure ps	479.7	480.0	474.7	550.0	469.2	469.2	469.2	469.2	469.2	469.2	2214.7	2214.7	20.0	15.6	14.7
Mole Flow lbmol/hr	63,523	7,796	69,187	5,808	44,373	6,887	37,487	2,699	1,102	15,522	18,164	61,067	54	36,278	91,148
Mass Flow lb/hr	1,612,600	247,900	1,860,480	104,624	1,141,680	124,107	1,017,580	30,622	12,506	176,128	798,320	2,686,500	968	1,043,850	2,482,890
Volume Flow cuft/hr	760,257	110,314	1,602,260	92,986	759,488	2,367	458,399	34,972	14,283	201,153	17,019	85,349	18	17,775,600	48,545,500
Enthalpy MMBtu/hr	-4,720.0	4.8	-5,690.3	-588.5	-4,005.7	-847.5	-3,367.0	-41.0	-16.8	-236.0	-3,145.3	-10,483.9	-6.6	19.0	-1,332.8
Mole Frac															
H2O	0.000	0.000	0.253	1.000	0.157	1.000	0.003	0.003	0.003	0.003	0.002	0.001	1.000	0.007	0.120
CO2	0.397	0.000	0.281	0.000	0.442	0.000	0.523	0.076	0.076	0.076	0.998	0.999	0.000	0.000	0.021
O2	0.000	0.950	0.000	0.000	0.000	0.000	0.000	0.000	0.000	0.000	0.000	0.000	0.000	0.207	0.024
N2	0.097	0.050	0.094	0.000	0.082	0.000	0.097	0.188	0.188	0.188	0.000	0.000	0.000	0.786	0.835
CH4	0.008	0.000	0.000	0.000	0.000	0.000	0.000	0.000	0.000	0.000	0.000	0.000	0.000	0.000	0.000
CO	0.138	0.000	0.253	0.000	0.020	0.000	0.024	0.046	0.046	0.046	0.000	0.000	0.000	0.000	0.000
COS	0.000	0.000	0.000	0.000	0.000	0.000	0.000	0.000	0.000	0.000	0.000	0.000	0.000	0.000	0.000
H2	0.349	0.000	0.117	0.000	0.299	0.000	0.354	0.687	0.687	0.687	0.000	0.000	0.000	0.000	0.000
H2S	0.000	0.000	0.000	0.000	0.000	0.000	0.000	0.000	0.000	0.000	0.000	0.000	0.000	0.000	0.000
HCL	0.000	0.000	0.000	0.000	0.000	0.000	0.000	0.000	0.000	0.000	0.000	0.000	0.000	0.000	0.000
S8	0.000	0.000	0.000	0.000	0.000	0.000	0.000	0.000	0.000	0.000	0.000	0.000	0.000	0.000	0.000
C2H4	0.001	0.000	0.000	0.000	0.000	0.000	0.000	0.000	0.000	0.000	0.000	0.000	0.000	0.000	0.000
C2H6	0.003	0.000	0.000	0.000	0.000	0.000	0.000	0.000	0.000	0.000	0.000	0.000	0.000	0.000	0.000
C3H6	0.001	0.000	0.000	0.000	0.000	0.000	0.000	0.000	0.000	0.000	0.000	0.000	0.000	0.000	0.000
C3H8	0.003	0.000	0.000	0.000	0.000	0.000	0.000	0.000	0.000	0.000	0.000	0.000	0.000	0.000	0.000
ISOBU-01	0.001	0.000	0.000	0.000	0.000	0.000	0.000	0.000	0.000	0.000	0.000	0.000	0.000	0.000	0.000
N-BUT-01	0.002	0.000	0.000	0.000	0.000	0.000	0.000	0.000	0.000	0.000	0.000	0.000	0.000	0.000	0.000
1-BUT-01	0.001	0.000	0.000	0.000	0.000	0.000	0.000	0.000	0.000	0.000	0.000	0.000	0.000	0.000	0.000
1-PEN-01	0.000	0.000	0.000	0.000	0.000	0.000	0.000	0.000	0.000	0.000	0.000	0.000	0.000	0.000	0.000
N-PEN-01	0.000	0.000	0.000	0.000	0.000	0.000	0.000	0.000	0.000	0.000	0.000	0.000	0.000	0.000	0.000
1-HEX-01	0.000	0.000	0.000	0.000	0.000	0.000	0.000	0.000	0.000	0.000	0.000	0.000	0.000	0.000	0.000
N-HEX-01	0.000	0.000	0.000	0.000	0.000	0.000	0.000	0.000	0.000	0.000	0.000	0.000	0.000	0.000	0.000
C7-C11	0.000	0.000	0.000	0.000	0.000	0.000	0.000	0.000	0.000	0.000	0.000	0.000	0.000	0.000	0.000
C12-C18	0.000	0.000	0.000	0.000	0.000	0.000	0.000	0.000	0.000	0.000	0.000	0.000	0.000	0.000	0.000
C19-C24	0.000	0.000	0.000	0.000	0.000	0.000	0.000	0.000	0.000	0.000	0.000	0.000	0.000	0.000	0.000
C25PL	0.000	0.000	0.000	0.000	0.000	0.000	0.000	0.000	0.000	0.000	0.000	0.000	0.000	0.000	0.000
ALCS	0.000	0.000	0.000	0.000	0.000	0.000	0.000	0.000	0.000	0.000	0.000	0.000	0.000	0.000	0.000
Solids Mass Flow lb/hr	1,612,600	247,900	1,860,480	104,624	1,141,680	124,107	1,017,580	30,622	12,506	176,128	798,320	2,686,500	968	1,043,850	2,482,890
Enthalpy MMBtu/hr	-4720.0	4.8	-5690.3	-588.5	-4005.7	-847.5	-3367.0	-41.0	-16.8	-236.0	-3145.3	-10483.9	-6.6	19.0	-1332.8
Pressure ps	479.7	480.0	474.7	550.0	469.2	469.2	469.2		469.2	469.2	2214.7	2214.7	20.0	15.6	14.7
Temperature F															
Pressure ps	479.7	480.0	474.7	550.0	469.2	469.2	469.2		469.2	469.2	2214.7	2214.7	20.0	15.6	14.7
Mass Flow lb/hr	0	0	0	0	0	0	0	0	0	0	0	0	0	0	0
Enthalpy MMBtu/hr															

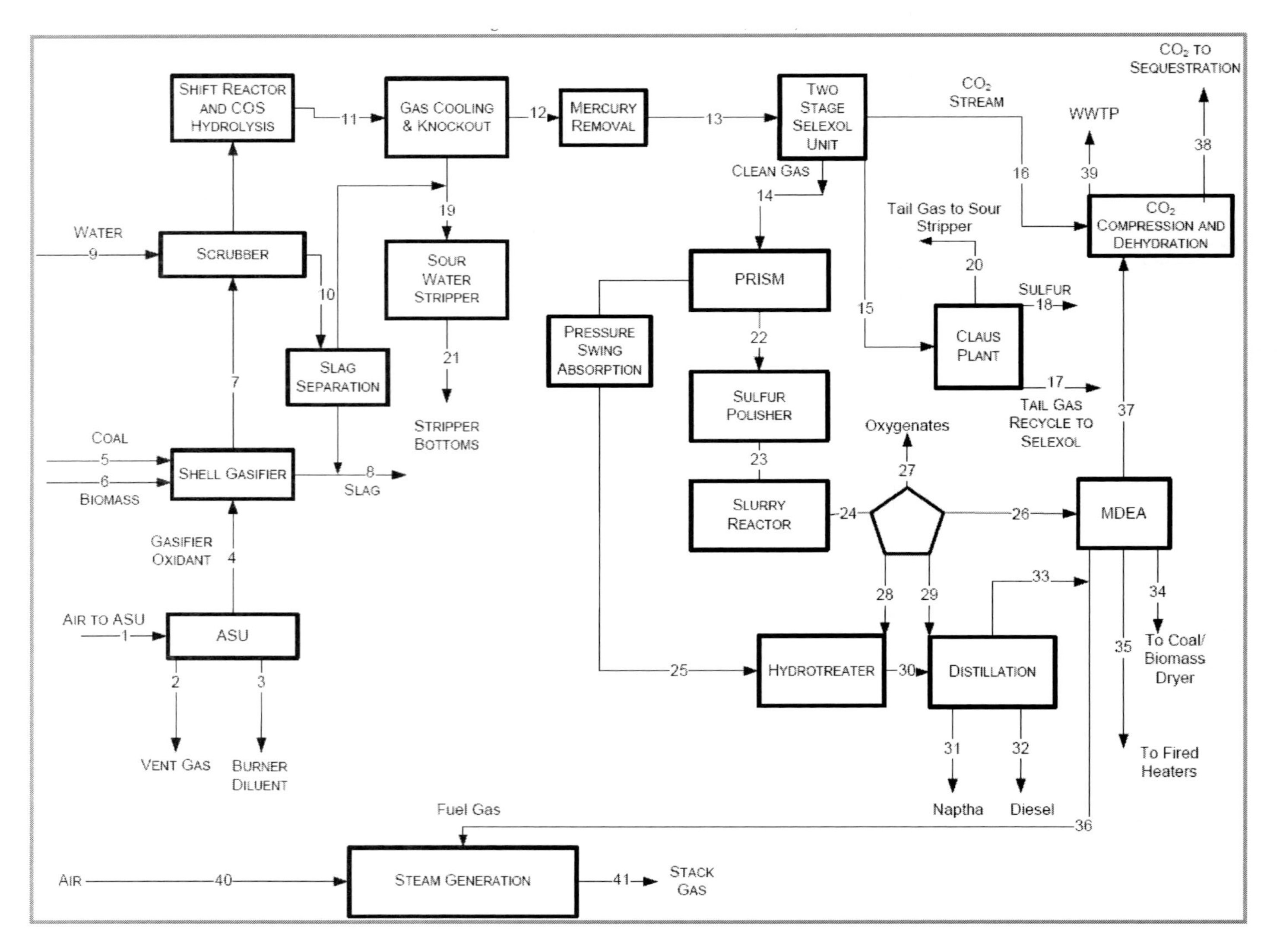

Figure D-4. 15wt% CBTL with CCS (Case 5).

Table D-6. 15wt% CBTL with CCS (Case 5)

Stream #	1	2	3	4	5	6	7	8	9	10
Temperature F	231.5	67.4	80.0	201.0	140.0	140.0	2565.7	100.0	280.0	411.8
Pressure psi	190.3	18.4	56.4	665.0	14.7	14.7	614.7	14.7	500.0	596.0
Mole Flow lbmol/hr	212,587	110,877	54,122	38,603	0	0	146,542	0	145,577	2,951
Mass Flow lb/hr	6,118,830	3,098,990	1,520,000	1,227,650	1,476,310	263,529	3,176,120	179,627	2,622,320	53,175
Volume Flow cuft/hr	8,911,420	37,463,500	5,860,000	409,165	0	0	7,925,200	0	53,321	1,222
Enthalpy MMBtu/hr	154.5	-163.4	4.2	29.3	-658.4	-593.8	-3,063.5	-8.2	-17,472.2	-344.7
Mole Frac										
H_2O	0.007	0.013	0.000	0.000	0.000	0.000	0.346	0.000	1.000	0.999
CO_2	0.000	0.000	0.000	0.000	0.000	0.000	0.033	0.000	0.000	0.000
O_2	0.207	0.018	0.006	0.950	0.000	0.000	0.000	0.000	0.000	0.000
N_2	0.786	0.969	0.992	0.050	0.000	0.000	0.020	0.000	0.000	0.000
CH_4	0.000	0.000	0.000	0.000	0.000	0.000	0.001	0.000	0.000	0.000
CO	0.000	0.000	0.000	0.000	0.000	0.000	0.326	0.000	0.000	0.000
COS	0.000	0.000	0.000	0.000	0.000	0.000	0.001	0.000	0.000	0.000
H_2	0.000	0.000	0.000	0.000	0.000	0.000	0.265	0.000	0.000	0.000
H_2S	0.000	0.000	0.000	0.000	0.000	0.000	0.008	0.000	0.000	0.000
HCL	0.000	0.000	0.000	0.000	0.000	0.000	0.001	0.000	0.000	0.000
S_8	0.000	0.000	0.000	0.000	0.000	0.000	0.000	0.000	0.000	0.000
C_2H_4	0.000	0.000	0.000	0.000	0.000	0.000	0.000	0.000	0.000	0.000
C_2H_6	0.000	0.000	0.000	0.000	0.000	0.000	0.000	0.000	0.000	0.000
C_3H_8	0.000	0.000	0.000	0.000	0.000	0.000	0.000	0.000	0.000	0.000
C_3H_6	0.000	0.000	0.000	0.000	0.000	0.000	0.000	0.000	0.000	0.000
ISOBU-01	0.000	0.000	0.000	0.000	0.000	0.000	0.000	0.000	0.000	0.000
N-BUT-01	0.000	0.000	0.000	0.000	0.000	0.000	0.000	0.000	0.000	0.000
1-BUT-01	0.000	0.000	0.000	0.000	0.000	0.000	0.000	0.000	0.000	0.000
1-PEN-01	0.000	0.000	0.000	0.000	0.000	0.000	0.000	0.000	0.000	0.000
N-PEN-01	0.000	0.000	0.000	0.000	0.000	0.000	0.000	0.000	0.000	0.000
1-HEX-01	0.000	0.000	0.000	0.000	0.000	0.000	0.000	0.000	0.000	0.000
N-HEX-01	0.000	0.000	0.000	0.000	0.000	0.000	0.000	0.000	0.000	0.000
C7-C11	0.000	0.000	0.000	0.000	0.000	0.000	0.000	0.000	0.000	0.000
C12-C18	0.000	0.000	0.000	0.000	0.000	0.000	0.000	0.000	0.000	0.000
C19-C24	0.000	0.000	0.000	0.000	0.000	0.000	0.000	0.000	0.000	0.000
C25PL	0.000	0.000	0.000	0.000	0.000	0.000	0.000	0.000	0.000	0.000
ALCS	0.000	0.000	0.000	0.000	0.000	0.000	0.000	0.000	0.000	0.000
Solids Mass Flow lb/hr	6,118,830	3,098,990	1,520,000	1,227,650	1,476,310	263,529	3,176,120	179,627	2,622,320	53,175
Enthalpy MMBtu/hr	154.5	-163.4	4.2	29.3	-658.4	-593.8	-3063.5	-8.2	-17472.2	-344.7
Pressure psi	190.3	18.4	56.4	665.0	14.7	14.7	614.7	14.7	500.0	596.0
Temperature F					140.0	140.0		100.0		
Pressure psi	190.3	18.4	56.4	665.0	14.7	14.7	614.7	14.7	500.0	596.0
Mass Flow lb/hr	0	0	0	0	1,476,310	263,529	0	179,627	0	0
Enthalpy MMBtu/hr					-658.4	-593.8		-8.2		

Stream #	11	12	13	14	15	16	17	18	19	20
Temperature F	383.5	100.0	98.7	100.0	100.0	175.8	110.0	320.0	298.5	110.3
Pressure psi	567.0	515.0	505.0	495.0	27.3	2214.7	549.5	25.0	515.0	75.0
Mole Flow lbmol/hr	292,168	109,924	109,924	132,661	2,193	26,212	919	155	124,900	1,607
Mass Flow lb/hr	5,748,560	3,542,850	3,542,850	2,247,490	63,611	1,153,560	35,356	39,870	2,263,570	28,960
Volume Flow cuft/hr	4,420,370	1,853,310	1,991,370	1,521,160	484,023	44,146	8,312	124	47,395	554
Enthalpy MMBtu/hr	-20,846.2	-8,614.3	-8,314.3	-3,200.0	-86.7	-4,483.5	-124.0	2.2	-14,352.7	-197.7
Mole Frac										
H_2O	0.419	0.002	0.002	0.000	0.012	0.000	0.003	0.000	0.998	1.000
CO_2	0.109	0.188	0.188	0.008	0.152	1.000	0.761	0.000	0.000	0.000
O_2	0.000	0.000	0.000	0.000	0.000	0.000	0.000	0.000	0.000	0.000
N_2	0.010	0.017	0.017	0.022	0.009	0.000	0.076	0.000	0.000	0.000
CH_4	0.000	0.000	0.000	0.001	0.000	0.000	0.001	0.000	0.000	0.000
CO	0.239	0.394	0.394	0.479	0.159	0.000	0.000	0.000	0.000	0.000
COS	0.000	0.000	0.000	0.000	0.000	0.000	0.001	0.000	0.000	0.000
H_2	0.229	0.392	0.392	0.490	0.099	0.000	0.081	0.000	0.000	0.000
H_2S	0.004	0.007	0.007	0.000	0.573	0.000	0.045	0.000	0.000	0.000
HCL	0.000	0.000	0.000	0.000	0.000	0.000	0.000	0.000	0.001	0.000
S_8	0.000	0.000	0.000	0.000	0.000	0.000	0.000	1.000	0.000	0.000
C_2H_4	0.000	0.000	0.000	0.000	0.000	0.000	0.000	0.000	0.000	0.000
C_2H_6	0.000	0.000	0.000	0.000	0.000	0.000	0.000	0.000	0.000	0.000
C_3H_8	0.000	0.000	0.000	0.000	0.000	0.000	0.000	0.000	0.000	0.000
C_3H_6	0.000	0.000	0.000	0.000	0.000	0.000	0.000	0.000	0.000	0.000
ISOBU-01	0.000	0.000	0.000	0.000	0.000	0.000	0.000	0.000	0.000	0.000
N-BUT-01	0.000	0.000	0.000	0.000	0.000	0.000	0.000	0.000	0.000	0.000
1-BUT-01	0.000	0.000	0.000	0.000	0.000	0.000	0.000	0.000	0.000	0.000
1-PEN-01	0.000	0.000	0.000	0.000	0.000	0.000	0.000	0.000	0.000	0.000
N-PEN-01	0.000	0.000	0.000	0.000	0.000	0.000	0.000	0.000	0.000	0.000
1-HEX-01	0.000	0.000	0.000	0.000	0.000	0.000	0.000	0.000	0.000	0.000
N-HEX-01	0.000	0.000	0.000	0.000	0.000	0.000	0.000	0.000	0.000	0.000
C7-C11	0.000	0.000	0.000	0.000	0.000	0.000	0.000	0.000	0.000	0.000
C12-C18	0.000	0.000	0.000	0.000	0.000	0.000	0.000	0.000	0.000	0.000
C19-C24	0.000	0.000	0.000	0.000	0.000	0.000	0.000	0.000	0.000	0.000
C25PL	0.000	0.000	0.000	0.000	0.000	0.000	0.000	0.000	0.000	0.000
ALCS	0.000	0.000	0.000	0.000	0.000	0.000	0.000	0.000	0.000	0.000
Solids Mass Flow lb/hr	5,748,560	3,542,850	3,542,850	2,247,490	63,611	1,153,560	35,356	39,870	2,263,570	28,960
Enthalpy MMBtu/hr	-20846.2	-8614.3	-8314.3	-3200.0	-86.7	-4483.5	-124.0	2.2	-14352.7	-197.7
Pressure psi	567.0	515.0	505.0	495.0	27.3	2214.7	549.5	25.0	515.0	75.0
Temperature F										
Pressure psi	567.0	515.0	505.0	495.0	27.3	2214.7	549.5	25.0	515.0	75.0
Mass Flow lb/hr	0	0	0	0	0	0	0	0	0	0
Enthalpy MMBtu/hr										

Table D-6. (Continued).

Stream #	21	22	23	24	25	26	27	28	29	30	31	32	33	34	35	36	37	38	39	40	41
Temperature F	247.0	897.8	495.0	408.3	100.0	100.0	100.0	453.3	103.0	323.3	77.0	77.0	100.0	103.0	100.0	90.8	105.0	141.3	82.8	250.0	270.0
Pressure psi	35.0	490.0	490.0	490.0	475.0	490.0	490.0	490.0	490.0	490.0	14.7	14.7	465.0	490.0	490.0	465.0	2214.7	2214.7	20.0	13.1	11.1
Mole Flow lbmol/hr	126,403	128,745	176,461	97,369	3,312	78,908	16,392	1,023	479	3,349	1,528	1,336	604	1,681	791	7,099	24,216	53,428	45	37,737	77,363
Mass Flow lb/hr	2,279,590	2,339,040	2,912,720	2,912,720	6,678	2,362,290	303,427	466,045	61,953	421,723	156,644	370,354	23,498	34,284	14,412	140,774	1,085,646	2,219,240	817	1,087,240	2,187,170
Volume Flow cuft/hr	152,754	3,368,900	3,733,160	1,807,460	42,353	330,902	5,541	13,115	1,425	158,388	3,676	8,322	6,200	22,682	8,661	68,651	22,800	63,178	16	22,061,800	55,183,300
Enthalpy MMBtu/hr	-15,139.0	-2,450.1	-3,440.8	-6,903.5	0.6	-4,920.7	-2,057.8	-395.3	-49.5	-366.2	-148.1	-324.5	-28.2	-23.2	-11.3	-124.6	-4,165.6	-8,579.3	-5.6	19.6	-1,188.7
Mole Frac																					
H2O	0.996	0.002	0.000	0.171	0.000	0.000	0.995	0.000	0.000	0.000	0.000	0.000	0.000	0.000	0.000	0.000	0.000	0.000	1.000	0.037	0.061
CO2	0.000	0.008	0.015	0.270	0.000	0.332	0.000	0.000	0.000	0.000	0.000	0.000	0.000	0.038	0.033	0.033	1.000	1.000	0.000	0.000	0.049
O2	0.000	0.000	0.000	0.000	0.000	0.000	0.000	0.000	0.000	0.000	0.000	0.000	0.000	0.000	0.000	0.000	0.000	0.000	0.000	0.207	0.126
N2	0.000	0.023	0.101	0.163	0.000	0.226	0.000	0.000	0.000	0.000	0.000	0.000	0.000	0.328	0.323	0.295	0.000	0.000	0.000	0.736	0.744
CH4	0.000	0.001	0.010	0.020	0.000	0.024	0.000	0.000	0.000	0.061	0.000	0.000	0.310	0.035	0.035	0.061	0.000	0.000	0.000	0.000	0.000
CO	0.000	0.495	0.412	0.322	0.000	0.114	0.000	0.000	0.000	0.000	0.000	0.000	0.000	0.164	0.164	0.149	0.000	0.000	0.000	0.000	0.000
COS	0.000	0.000	0.000	0.000	0.000	0.000	0.000	0.000	0.000	0.000	0.000	0.000	0.000	0.000	0.000	0.000	0.000	0.000	0.000	0.000	0.000
H2	0.000	0.473	0.453	0.226	1.000	0.275	0.000	0.000	0.000	0.000	0.000	0.000	0.000	0.402	0.402	0.364	0.000	0.000	0.000	0.000	0.000
H2S	0.000	0.000	0.000	0.000	0.000	0.000	0.000	0.000	0.000	0.000	0.000	0.000	0.000	0.000	0.000	0.000	0.000	0.000	0.000	0.000	0.000
HCL	0.001	0.000	0.000	0.000	0.000	0.000	0.000	0.000	0.000	0.000	0.000	0.000	0.000	0.000	0.000	0.000	0.000	0.000	0.000	0.000	0.000
38	0.000	0.000	0.000	0.000	0.000	0.000	0.000	0.000	0.000	0.000	0.000	0.000	0.000	0.000	0.000	0.000	0.000	0.000	0.000	0.000	0.000
C2H4	0.000	0.000	0.003	0.008	0.000	0.008	0.000	0.000	0.000	0.000	0.000	0.000	0.000	0.011	0.011	0.010	0.000	0.000	0.000	0.000	0.000
C2H6	0.000	0.000	0.001	0.002	0.000	0.002	0.000	0.000	0.000	0.043	0.000	0.000	0.230	0.003	0.003	0.025	0.000	0.000	0.000	0.000	0.000
C3H6	0.000	0.000	0.002	0.005	0.000	0.007	0.000	0.000	0.000	0.000	0.000	0.000	0.000	0.008	0.008	0.008	0.000	0.000	0.000	0.000	0.000
C3H8	0.000	0.000	0.001	0.002	0.000	0.002	0.000	0.000	0.000	0.043	0.000	0.000	0.230	0.003	0.003	0.024	0.000	0.000	0.000	0.000	0.000
ISOBU-01	0.000	0.000	0.000	0.000	0.000	0.000	0.000	0.000	0.000	0.023	0.000	0.000	0.115	0.003	0.003	0.011	0.000	0.000	0.000	0.000	0.000
N-BUT-01	0.000	0.000	0.001	0.001	0.000	0.002	0.000	0.000	0.000	0.023	0.000	0.000	0.115	0.003	0.003	0.013	0.000	0.000	0.000	0.000	0.000
1-BUT-01	0.000	0.000	0.002	0.005	0.000	0.006	0.000	0.000	0.000	0.000	0.000	0.000	0.000	0.009	0.009	0.007	0.000	0.000	0.000	0.000	0.000
1-PEN-01	0.000	0.000	0.000	0.001	0.000	0.000	0.000	0.000	0.145	0.000	0.045	0.000	0.000	0.000	0.000	0.000	0.000	0.000	0.000	0.000	0.000
N-PEN-01	0.000	0.000	0.000	0.000	0.000	0.000	0.000	0.000	0.015	0.043	0.105	0.000	0.000	0.000	0.000	0.000	0.000	0.000	0.000	0.000	0.000
1-HEX-01	0.000	0.000	0.000	0.001	0.000	0.000	0.000	0.000	0.124	0.000	0.039	0.000	0.000	0.000	0.000	0.000	0.000	0.000	0.000	0.000	0.000
N-HEX-01	0.000	0.000	0.000	0.000	0.000	0.000	0.000	0.000	0.014	0.043	0.105	0.000	0.000	0.000	0.000	0.000	0.000	0.000	0.000	0.000	0.000
C7-C11	0.000	0.000	0.000	0.002	0.000	0.000	0.000	0.000	0.473	0.264	0.706	0.000	0.000	0.000	0.000	0.000	0.000	0.000	0.000	0.000	0.000
C12-C18	0.000	0.000	0.000	0.001	0.000	0.000	0.000	0.000	0.223	0.344	0.000	0.772	0.000	0.000	0.000	0.000	0.000	0.000	0.000	0.000	0.000
C19-C24	0.000	0.000	0.000	0.001	0.000	0.000	0.000	0.066	0.000	0.111	0.000	0.228	0.000	0.000	0.000	0.000	0.000	0.000	0.000	0.000	0.000
C25PL	0.000	0.000	0.000	0.010	0.000	0.000	0.000	0.934	0.000	0.000	0.000	0.000	0.000	0.000	0.000	0.000	0.000	0.000	0.000	0.000	0.000
ALCS	0.000	0.000	0.000	0.001	0.000	0.000	0.005	0.000	0.000	0.000	0.000	0.000	0.000	0.000	0.000	0.000	0.000	0.000	0.000	0.000	0.000
Solids Mass Flow lb/hr	2,279,590	2,339,040	2,912,720	2,912,720	6,678	2,362,290	303,427	466,045	61,953	421,723	156,644	370,354	23,498	34,284	14,412	140,774	1,070,000	2,219,240	817	1,087,240	2,187,170
Enthalpy MMBtu/hr	-15139.0	-2450.1	-3440.8	-6903.5	0.6	-4920.7	-2057.8	-395.3	-49.5	-366.2	-148.1	-324.5	-28.2	-23.2	-11.3	-124.6	-4165.6	-8579.3	-5.6	19.6	-1188.7
Pressure psi	35.0	490.0	490.0	490.0	475.0	490.0	490.0	490.0	490.0	490.0	14.7	14.7	465.0		490.0	465.0	2214.7	2214.7	20.0	13.1	11.1
Temperature F																					
Pressure psi	35.0	490.0	490.0	490.0	475.0	490.0	490.0	490.0	490.0	490.0	14.7	14.7	465.0		490.0	465.0	2214.7	2214.7	20.0	13.1	11.1
Mass Flow lb/hr	0	0	0	0	0	0	0	0	0	0	0	0	0	0	0	0	0	0	0	0	0
Enthalpy MMBtu/hr																					

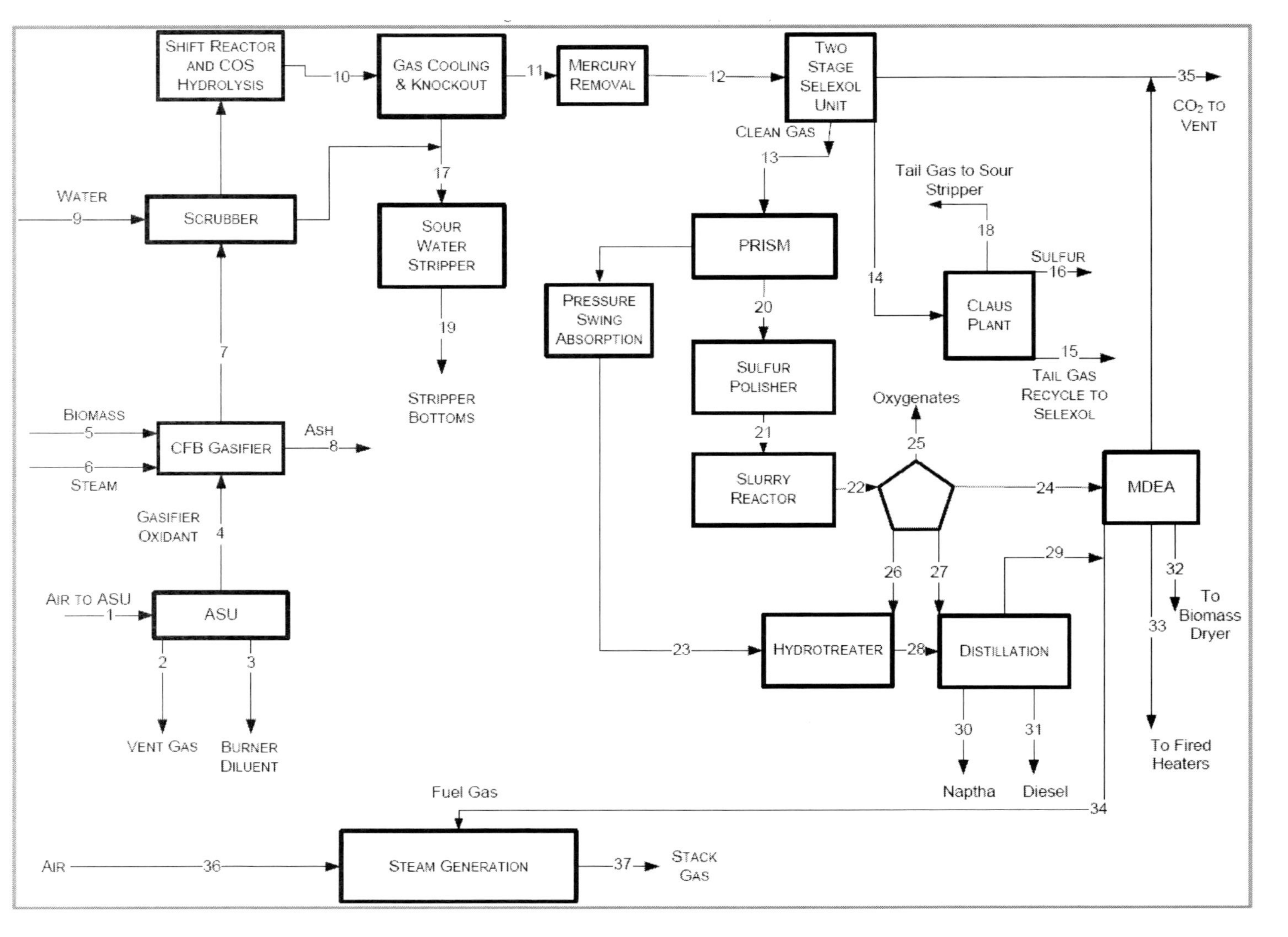

Figure D-5. BTL without CCS (Case 9).

Table D-6. without CCS (Case 9)

Stream #	1	2	3	4	5	6	7	8	9	10	11	12	13	14	15	16	17	18
Temperature F	281.5	57.6	90.0	180.7	140.0	466.0	2000.0	100.0	390.0	400.9	100.0	99.7	100.0	100.0	110.0	320.0	303.3	110.5
Pressure psi	190.3	15.4	56.4	500.0	14.7	500.0	449.7	14.7	460.0	410.2	372.7	362.7	350.0	27.0	324.5	25.0	372.7	75.0
Mole Flow lbmol/hr	24,973	11,619	8,211	5,098	0	8,602	31,582	0	30,384	61,346	21,506	21,506	15,208	148	121	1	40,398	61
Mass Flow lb/hr	718,545	324,727	230,283	162,103	322,432	154,966	667,360	20,384	547,382	1,203,580	485,531	485,531	223,313	4,370	5,164	275	728,095	1,094
Volume Flow cuft/hr	1,046,930	3,921,040	358,000	69,568	0	143,134	1,862,850	0	12,336	1,259,520	340,302	349,603	261,915	30,669	2,052	1	15,372	21
Enthalpy MMBtu/hr	18.2	-19.3	0.6	3.2	-725.1	-373.1	-1,841.0	-3.0	-3,564.7	-5,372.3	-1,340.2	-1,340.2	-352.7	-13.9	-19.2	0.0	-4,813.0	-7.5
Mole Frac																		
H2O	0.007	0.014	0.000	0.000	0.000	1.000	0.336	0.000	1.000	0.650	0.003	0.003	0.000	0.095	0.004	0.000	0.999	1.000
CO2	0.000	0.000	0.000	0.000	0.000	0.000	0.171	0.000	0.000	0.097	0.275	0.275	0.013	0.416	0.936	0.000	0.000	0.000
O2	0.207	0.019	0.008	0.950	0.000	0.000	0.000	0.000	0.000	0.000	0.000	0.000	0.000	0.000	0.000	0.000	0.000	0.000
N2	0.786	0.967	0.992	0.050	0.000	0.000	0.011	0.000	0.000	0.006	0.016	0.016	0.023	0.005	0.049	0.000	0.000	0.000
CH4	0.000	0.000	0.000	0.000	0.000	0.000	0.001	0.000	0.000	0.000	0.001	0.001	0.001	0.002	0.002	0.000	0.000	0.000
CO	0.000	0.000	0.000	0.000	0.000	0.000	0.240	0.000	0.000	0.115	0.328	0.328	0.442	0.248	0.000	0.000	0.000	0.000
H2	0.000	0.000	0.000	0.000	0.000	0.000	0.240	0.000	0.000	0.132	0.377	0.377	0.520	0.176	0.006	0.000	0.000	0.000
H2S	0.000	0.000	0.000	0.000	0.000	0.000	0.000	0.000	0.000	0.000	0.000	0.000	0.000	0.059	0.003	0.000	0.000	0.000
S8	0.000	0.000	0.000	0.000	0.000	0.000	0.000	0.000	0.000	0.000	0.000	0.000	0.000	0.000	0.000	1.000	0.000	0.000
C2H4	0.000	0.000	0.000	0.000	0.000	0.000	0.000	0.000	0.000	0.000	0.000	0.000	0.000	0.000	0.000	0.000	0.000	0.000
C2H6	0.000	0.000	0.000	0.000	0.000	0.000	0.000	0.000	0.000	0.000	0.000	0.000	0.000	0.000	0.000	0.000	0.000	0.000
C3H6	0.000	0.000	0.000	0.000	0.000	0.000	0.000	0.000	0.000	0.000	0.000	0.000	0.000	0.000	0.000	0.000	0.000	0.000
C3H8	0.000	0.000	0.000	0.000	0.000	0.000	0.000	0.000	0.000	0.000	0.000	0.000	0.000	0.000	0.000	0.000	0.000	0.000
ISOBU-01	0.000	0.000	0.000	0.000	0.000	0.000	0.000	0.000	0.000	0.000	0.000	0.000	0.000	0.000	0.000	0.000	0.000	0.000
N-BUT-01	0.000	0.000	0.000	0.000	0.000	0.000	0.000	0.000	0.000	0.000	0.000	0.000	0.000	0.000	0.000	0.000	0.000	0.000
1-BUT-01	0.000	0.000	0.000	0.000	0.000	0.000	0.000	0.000	0.000	0.000	0.000	0.000	0.000	0.000	0.000	0.000	0.000	0.000
1-PEN-01	0.000	0.000	0.000	0.000	0.000	0.000	0.000	0.000	0.000	0.000	0.000	0.000	0.000	0.000	0.000	0.000	0.000	0.000
N-PEN-01	0.000	0.000	0.000	0.000	0.000	0.000	0.000	0.000	0.000	0.000	0.000	0.000	0.000	0.000	0.000	0.000	0.000	0.000
1-HEX-01	0.000	0.000	0.000	0.000	0.000	0.000	0.000	0.000	0.000	0.000	0.000	0.000	0.000	0.000	0.000	0.000	0.000	0.000
N-HEX-01	0.000	0.000	0.000	0.000	0.000	0.000	0.000	0.000	0.000	0.000	0.000	0.000	0.000	0.000	0.000	0.000	0.000	0.000
C7-C11	0.000	0.000	0.000	0.000	0.000	0.000	0.000	0.000	0.000	0.000	0.000	0.000	0.000	0.000	0.000	0.000	0.000	0.000
C12-C18	0.000	0.000	0.000	0.000	0.000	0.000	0.000	0.000	0.000	0.000	0.000	0.000	0.000	0.000	0.000	0.000	0.000	0.000
C19-C24	0.000	0.000	0.000	0.000	0.000	0.000	0.000	0.000	0.000	0.000	0.000	0.000	0.000	0.000	0.000	0.000	0.000	0.000
C25PL	0.000	0.000	0.000	0.000	0.000	0.000	0.000	0.000	0.000	0.000	0.000	0.000	0.000	0.000	0.000	0.000	0.000	0.000
ALCS	0.000	0.000	0.000	0.000	0.000	0.000	0.000	0.000	0.000	0.000	0.000	0.000	0.000	0.000	0.000	0.000	0.000	0.000
Solids Mass Flow lb/hr	718,545	324,727	230,000	162,103	322,432	154,966	667,360	20,384	547,382	1,203,580	485,531	485,531	223,313	4,370	5,164	275	728,095	1,094
Enthalpy MMBtu/hr	18.2	-19.3	0.6	3.2	-725.1	-373.1	-1841.0	-3.0	-3564.7	-5372.3	-1340.2	-1340.2	-352.7	-13.9	-19.2	0.0	-4813.0	-7.5
Pressure psi	190.3	15.4	56.4	500.0	14.7	500.0	449.7	14.7	460.0	410.2	372.7	362.7	350.0	27.0	324.5	25.0	372.7	75.0
Temperature F					140.0			100.0										
Pressure psi	190.3	15.4	56.4	500.0	14.7	500.0	449.7	14.7	460.0	410.2	372.7	362.7	350.0	27.0	324.5	25.0	372.7	75.0
Mass Flow lb/hr	0	0	0	0	322,432	0	0	20,384	0	0	0	0	0	0	0	0	0	0
Enthalpy MMBtu/hr					-725.1			-3.0										

Stream #	19	20	21	22	23	24	25	26	27	28	29	30	31	32	33	34	35	36	37
Temperature F	247.0	432.4	174.4	400.7	100.0	100.0	100.0	452.3	100.0	330.9	100.0	77.0	77.0	100.0	100.0	98.2	48.6	260.0	270.0
Pressure psi	35.0	345.0	345.0	295.0	330.0	295.0	295.0	295.0	295.0	295.0	295.0	14.7	14.7	295.0	295.0	295.0	20.0	16.1	16.1
Mole Flow lbmol/hr	40,435	14,795	17,602	9,476	331	7,602	1,824	102	49	335	87	153	163	312	76	2,183	7,323	13,188	18,135
Mass Flow lb/hr	728,450	222,484	258,250	268,250	658	170,273	33,140	43,626	6,211	49,294	2,355	18,024	37,125	4,129	1,006	30,371	316,184	378,318	508,413
Volume Flow cuft/hr	14,857	413,911	347,147	292,369	6,075	149,853	633	1,316	143	16,738	1,067	383	901	6,270	1,556	43,375	1,990,330	6,246,980	6,934,650
Enthalpy MMBtu/hr	-4,884.5	-317.7	-387.5	-681.0	0.1	-476.1	-224.9	-39.7	-5.0	-36.8	-2.3	-14.6	-32.4	-5.1	-1.2	-36.9	-1,208.7	6.9	-283.1
Mole Frac																			
H2O	1.000	0.000	0.000	0.192	0.000	0.000	0.996	0.000	0.000	0.000	0.000	0.000	0.000	0.000	0.000	0.000	0.003	0.007	0.094
CO2	0.000	0.014	0.017	0.263	0.000	0.333	0.000	0.000	0.000	0.000	0.000	0.000	0.000	0.036	0.036	0.035	0.966	0.000	0.043
O2	0.000	0.000	0.000	0.000	0.000	0.000	0.000	0.000	0.000	0.000	0.000	0.000	0.000	0.000	0.000	0.000	0.000	0.207	0.031
N2	0.000	0.024	0.042	0.077	0.000	0.097	0.000	0.000	0.000	0.000	0.000	0.000	0.000	0.140	0.140	0.138	0.000	0.793	0.732
CH4	0.000	0.001	0.004	0.006	0.000	0.012	0.000	0.000	0.000	0.032	0.313	0.000	0.000	0.017	0.017	0.028	0.000	0.000	0.000
CO	0.000	0.454	0.418	0.112	0.000	0.142	0.000	0.000	0.000	0.000	0.000	0.000	0.000	0.204	0.204	0.188	0.020	0.000	0.000
H2	0.000	0.507	0.520	0.323	1.000	0.408	0.000	0.000	0.000	0.000	0.000	0.000	0.000	0.590	0.590	0.572	0.010	0.000	0.000
H2S	0.000	0.000	0.000	0.000	0.000	0.000	0.000	0.000	0.000	0.000	0.000	0.000	0.000	0.000	0.000	0.000	0.000	0.000	0.000
S8	0.000	0.000	0.000	0.000	0.000	0.000	0.000	0.000	0.000	0.000	0.000	0.000	0.000	0.000	0.000	0.000	0.000	0.000	0.000
C2H4	0.000	0.000	0.001	0.002	0.000	0.003	0.000	0.000	0.000	0.000	0.000	0.000	0.000	0.004	0.004	0.004	0.000	0.000	0.000
C2H6	0.000	0.000	0.000	0.001	0.000	0.001	0.000	0.000	0.000	0.048	0.233	0.000	0.000	0.001	0.001	0.008	0.000	0.000	0.000
C3H6	0.000	0.000	0.000	0.002	0.000	0.002	0.000	0.000	0.000	0.000	0.000	0.000	0.000	0.003	0.003	0.003	0.000	0.000	0.000
C3H8	0.000	0.000	0.000	0.001	0.000	0.001	0.000	0.000	0.000	0.048	0.233	0.000	0.000	0.001	0.001	0.008	0.000	0.000	0.000
ISOBU-01	0.000	0.000	0.000	0.000	0.000	0.000	0.000	0.000	0.000	0.023	0.115	0.000	0.000	0.000	0.000	0.004	0.000	0.000	0.000
N-BUT-01	0.000	0.000	0.000	0.000	0.000	0.001	0.000	0.000	0.000	0.023	0.115	0.000	0.000	0.001	0.001	0.004	0.000	0.000	0.000
1-BUT-01	0.000	0.000	0.000	0.001	0.000	0.002	0.000	0.000	0.000	0.000	0.000	0.000	0.000	0.003	0.003	0.003	0.000	0.000	0.000
1-PEN-01	0.000	0.000	0.000	0.001	0.000	0.000	0.000	0.000	0.145	0.000	0.000	0.045	0.000	0.000	0.000	0.000	0.000	0.000	0.000
N-PEN-01	0.000	0.000	0.000	0.000	0.000	0.000	0.000	0.000	0.316	0.048	0.000	0.105	0.000	0.000	0.000	0.000	0.000	0.000	0.000
1-HEX-01	0.000	0.000	0.000	0.001	0.000	0.000	0.000	0.000	0.125	0.000	0.000	0.039	0.000	0.000	0.000	0.000	0.000	0.000	0.000
N-HEX-01	0.000	0.000	0.000	0.000	0.000	0.000	0.000	0.000	0.314	0.048	0.000	0.105	0.000	0.000	0.000	0.000	0.000	0.000	0.000
C7-C11	0.000	0.000	0.000	0.002	0.000	0.000	0.000	0.000	0.473	0.256	0.000	0.706	0.000	0.000	0.000	0.000	0.000	0.000	0.000
C12-C18	0.000	0.000	0.000	0.001	0.000	0.000	0.000	0.000	0.228	0.344	0.000	0.000	0.772	0.000	0.000	0.000	0.000	0.000	0.000
C19-C24	0.000	0.000	0.000	0.001	0.000	0.000	0.000	0.066	0.000	0.111	0.000	0.000	0.228	0.000	0.000	0.000	0.000	0.000	0.000
C25PL	0.000	0.000	0.000	0.010	0.000	0.000	0.000	0.934	0.000	0.000	0.000	0.000	0.000	0.000	0.000	0.000	0.000	0.000	0.000
ALGS	0.000	0.000	0.000	0.001	0.000	0.000	0.004	0.000	0.000	0.000	0.000	0.000	0.000	0.000	0.000	0.000	0.000	0.000	0.000
Solids Mass Flow lb/hr	728,450	222,484	258,250	268,250	658	170,273	33,140	43,626	6,211	49,294	2,355	18,024	37,125	4,129	1,006	30,371	316,184	378,318	508,413
Enthalpy MMBtu/hr	-4884.5	-317.7	-387.5	-681.0	0.1	-476.1	-224.9	-39.7	-5.0	-36.8	-2.3	-14.6	-32.4	-5.1	-1.2	-36.9	-1208.7	6.9	-283.1
Pressure psi	35.0	345.0	345.0	295.0	330.0	295.0	295.0	295.0	295.0	295.0	295.0	14.7	14.7		295.0	295.0	20.0	16.1	16.1
Temperature F																			
Pressure psi	35.0	345.0	345.0	295.0	330.0	295.0	295.0	295.0	295.0	295.0	295.0	14.7	14.7		295.0	295.0	20.0	16.1	16.1
Mass Flow lb/hr	0	0	0	0	0	0	0	0	0	0	0	0	0	0	0	0	0	0	0
Enthalpy MMBtu/hr																			

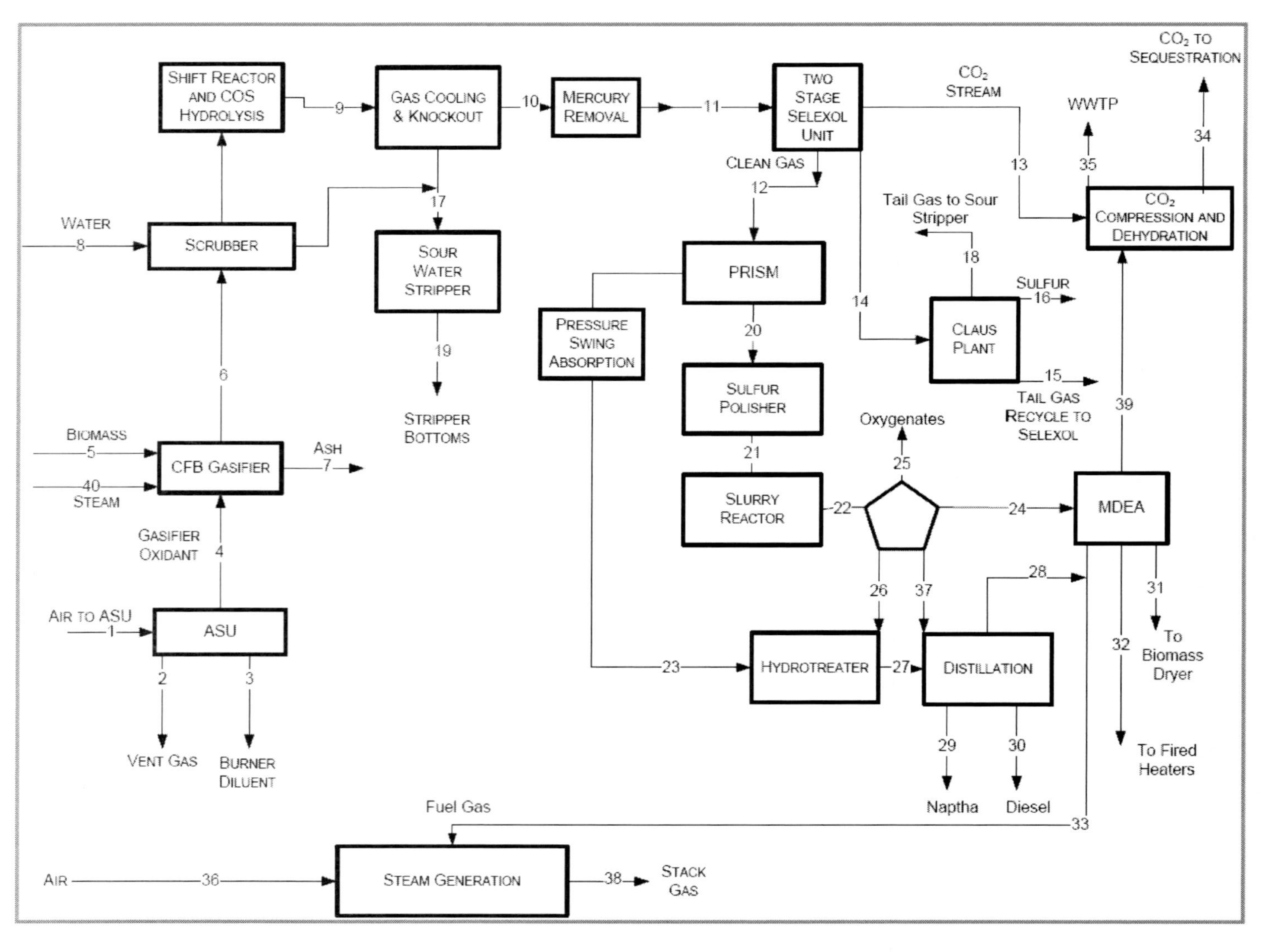

Figure D-6. BTL with CCS (Case 10).

Table D-7. BTL with CCS (Case 10)

Viso Stream #	1	2	3	4	5	6	7	8	9	10	11	12	13	14	15	16	17	18	19	20
Temperature F	281.8	57.8	90.0	180.7	140.0	2000.0	100.0	390.0	400.0	100.0	99.7	100.0	175.6	100.0	110.0	320.0	303.1	110.5	247.0	456.7
Pressure psi	190.3	16.4	56.4	500.0	14.7	449.7	14.7	460.0	410.2	372.7	382.7	350.0	2214.7	27.0	324.6	25.0	372.7	75.0	35.0	345.0
Mole Flow lbmol/hr	25,288	11,763	8,312	5,159	0	31,973	0	30,759	62,104	21,232	21,632	15,412	4,970	151	123	1	40,735	62	40,774	14,998
Mass Flow lb/hr	727,033	328,478	233,101	164,022	328,528	875,564	20,843	564,123	1,213,380	494,366	464,368	222,566	216,739	4,470	5,289	279	734,191	1,115	734,561	221,718
Volume Flow cuft/hr	1,059,190	3,966,300	869,000	70,391	0	1,835,900	0	12,488	1,262,390	348,329	358,412	285,441	6,371	31,344	2,093	1	15,409	21	14,381	431,017
Enthalpy MMBtu/hr	18.4	-19.5	0.7	3.3	-734.3	-1,863.2	-3.1	-3,603.8	-5,449.7	-1,375.6	-1,375.8	-351.1	-650.1	-14.3	-19.6	0.0	-4,853.5	-7.3	-4,906.2	-313.0
Mole Frac																				
H2O	0.007	0.014	0.000	0.000	0.000	0.338	0.000	1.000	0.347	0.003	0.003	0.000	0.000	0.093	0.004	0.000	0.999	1.000	1.000	0.000
CO2	0.000	0.000	0.000	0.000	0.000	0.171	0.000	0.000	0.099	0.280	0.280	0.014	1.000	0.423	0.936	0.000	0.000	0.000	0.000	0.014
O2	0.207	0.019	0.008	0.950	0.000	0.000	0.000	0.000	0.000	0.000	0.000	0.000	0.000	0.000	0.000	0.000	0.000	0.000	0.000	0.000
N2	0.786	0.967	0.992	0.050	0.000	0.011	0.000	0.000	0.006	0.016	0.016	0.023	0.000	0.005	0.048	0.000	0.000	0.000	0.000	0.023
CH4	0.000	0.000	0.000	0.000	0.000	0.001	0.000	0.000	0.000	0.001	0.001	0.001	0.000	0.002	0.002	0.000	0.000	0.000	0.000	0.001
CO	0.000	0.000	0.000	0.000	0.000	0.240	0.000	0.000	0.113	0.319	0.319	0.432	0.000	0.241	0.000	0.000	0.000	0.000	0.000	0.444
COS	0.000	0.000	0.000	0.000	0.000	0.000	0.000	0.000	0.000	0.000	0.000	0.000	0.000	0.000	0.000	0.000	0.000	0.000	0.000	0.000
H2	0.000	0.000	0.000	0.000	0.000	0.240	0.000	0.000	0.135	0.361	0.381	0.530	0.000	0.178	0.006	0.000	0.000	0.000	0.000	0.517
H2S	0.000	0.000	0.000	0.000	0.000	0.000	0.000	0.000	0.000	0.000	0.000	0.000	0.000	0.058	0.003	0.000	0.000	0.000	0.000	0.000
S8	0.000	0.000	0.000	0.000	0.000	0.000	0.000	0.000	0.000	0.000	0.000	0.000	0.000	0.000	0.000	1.000	0.000	0.000	0.000	0.000
C2H4	0.000	0.000	0.000	0.000	0.000	0.000	0.000	0.000	0.000	0.000	0.000	0.000	0.000	0.000	0.000	0.000	0.000	0.000	0.000	0.000
C2H6	0.000	0.000	0.000	0.000	0.000	0.000	0.000	0.000	0.000	0.000	0.000	0.000	0.000	0.000	0.000	0.000	0.000	0.000	0.000	0.000
C3H6	0.000	0.000	0.000	0.000	0.000	0.000	0.000	0.000	0.000	0.000	0.000	0.000	0.000	0.000	0.000	0.000	0.000	0.000	0.000	0.000
C3H8	0.000	0.000	0.000	0.000	0.000	0.000	0.000	0.000	0.000	0.000	0.000	0.000	0.000	0.000	0.000	0.000	0.000	0.000	0.000	0.000
ISOBU-01	0.000	0.000	0.000	0.000	0.000	0.000	0.000	0.000	0.000	0.000	0.000	0.000	0.000	0.000	0.000	0.000	0.000	0.000	0.000	0.000
N-BUT-01	0.000	0.000	0.000	0.000	0.000	0.000	0.000	0.000	0.000	0.000	0.000	0.000	0.000	0.000	0.000	0.000	0.000	0.000	0.000	0.000
1-BUT-01	0.000	0.000	0.000	0.000	0.000	0.000	0.000	0.000	0.000	0.000	0.000	0.000	0.000	0.000	0.000	0.000	0.000	0.000	0.000	0.000
1-PEN-01	0.000	0.000	0.000	0.000	0.000	0.000	0.000	0.000	0.000	0.000	0.000	0.000	0.000	0.000	0.000	0.000	0.000	0.000	0.000	0.000
N-PEN-01	0.000	0.000	0.000	0.000	0.000	0.000	0.000	0.000	0.000	0.000	0.000	0.000	0.000	0.000	0.000	0.000	0.000	0.000	0.000	0.000
1-HEX-01	0.000	0.000	0.000	0.000	0.000	0.000	0.000	0.000	0.000	0.000	0.000	0.000	0.000	0.000	0.000	0.000	0.000	0.000	0.000	0.000
N-HEX-01	0.000	0.000	0.000	0.000	0.000	0.000	0.000	0.000	0.000	0.000	0.000	0.000	0.000	0.000	0.000	0.000	0.000	0.000	0.000	0.000
C7-C11	0.000	0.000	0.000	0.000	0.000	0.000	0.000	0.000	0.000	0.000	0.000	0.000	0.000	0.000	0.000	0.000	0.000	0.000	0.000	0.000
C12-C18	0.000	0.000	0.000	0.000	0.000	0.000	0.000	0.000	0.000	0.000	0.000	0.000	0.000	0.000	0.000	0.000	0.000	0.000	0.000	0.000
C19-C24	0.000	0.000	0.000	0.000	0.000	0.000	0.000	0.000	0.000	0.000	0.000	0.000	0.000	0.000	0.000	0.000	0.000	0.000	0.000	0.000
C25FL	0.000	0.000	0.000	0.000	0.000	0.000	0.000	0.000	0.000	0.000	0.000	0.000	0.000	0.000	0.000	0.000	0.000	0.000	0.000	0.000
ALCS	0.000	0.000	0.000	0.000	0.000	0.000	0.000	0.000	0.000	0.000	0.000	0.000	0.000	0.000	0.000	0.000	0.000	0.000	0.000	0.000
Solids Mass Flow lb/hr	727,033	328,478	233,000	164,022	328,528	875,564	20,843	564,123	1,213,380	494,366	464,368	222,566	216,738	4,470	5,289	279	734,191	1,115	734,561	221,718
Enthalpy MMBtu/hr	18.4	-19.5	0.7	3.3	-734.3	-1863.2	-3.1	-3603.8	-5449.7	-1375.6	-1375.8	-351.1	-650.1	-14.3	-19.6	0.0	-4853.5	-7.3	-4906.2	-313.0
Pressure psi	190.3	16.4	56.4	500.0	14.7	449.7	14.7	460.0	410.2	372.7	362.7	350.0	2214.7	27.0	324.6	25.0	372.7	75.0	35.0	345.0
Temperature F					140.0		100.0													
Pressure psi	190.3	16.4	56.4	500.0	14.7	449.7	14.7	460.0	410.2	372.7	362.7	350.0	2214.7	27.0	324.5	25.0	372.7	75.0	35.0	345.0
Mass Flow lb/hr	0	0	0	0	328,528	0	20,843	0	0	0	0	0	0	0	0	0	0	0	0	0
Enthalpy MMBtu/hr					-734.3		-3.1													

Table D-7. (Continued).

Visio Stream #	21	22	23	24	25	26	27	28	29	30	31	32	33	34	35	36	37	38	39	40
Temperature F	136.2	399.1	100.0	100.0	100.0	452.0	330.5	100.0	77.0	77.0	100.0	100.0	98.5	152.1	77.3	250.0	100.0	270.0	105.0	466.0
Pressure psi	295.0	295.0	330.0	295.0	295.0	295.0	295.0	295.0	14.7	14.7	295.0	295.0	295.0	2200.0	23.0	16.1	295.0	16.1	2200.0	500.0
Mole Flow lbmol/hr	17,131	8,226	331	7,241	1,839	102	335	66	152	164	316	75	2,748	7,232	24	17,016	48	22,313	2,262	8,734
Mass Flow lb/hr	249,961	249,961	668	162,341	33,397	48,347	49,015	2,342	15,233	36,915	3,386	946	36,013	318,262	441	489,676	6,175	626,335	99,523	156,301
Volume Flow cuft/hr	410,917	293,263	6,276	144,380	642	1,311	15,328	1,081	386	603	6,444	1,538	55,324	10,067	8	6,078,840	143	10,837,330	2,149	149,388
Enthalpy MMBtu/hr	-377.6	-672.7	0.1	-464.4	-226.7	-39.6	-36.8	-2.8	-14.6	-32.5	-6.2	-1.2	-46.8	-1,242.0	-3.0	8.8	-4.9	-350.0	-391.9	-863.4
Mole Frac																				
H2O	0.000	0.198	0.000	0.000	0.996	0.000	0.000	0.000	0.000	0.000	0.000	0.000	0.000	0.000	1.000	0.007	0.000	0.096	0.000	1.000
CO2	0.016	0.265	0.000	0.338	0.000	0.000	0.000	0.000	0.000	0.000	0.387	0.037	0.386	1.000	0.000	0.000	0.000	0.042	1.000	0.000
O2	0.000	0.000	0.000	0.000	0.000	0.000	0.000	0.000	0.000	0.000	0.000	0.000	0.000	0.000	0.000	0.227	0.000	0.089	0.000	0.000
N2	0.034	0.062	0.000	0.079	0.000	0.000	0.000	0.000	0.000	0.000	0.115	0.115	0.113	0.000	0.000	0.756	0.000	0.773	0.000	0.000
CH4	0.003	0.008	0.000	0.010	0.000	0.000	0.081	0.310	0.000	0.000	0.014	0.014	0.021	0.000	0.000	0.000	0.000	0.000	0.000	0.000
CO	0.420	0.112	0.000	0.143	0.000	0.000	0.000	0.000	0.000	0.000	0.208	0.208	0.203	0.000	0.000	0.000	0.000	0.000	0.000	0.000
COS	0.000	0.000	0.000	0.000	0.000	0.000	0.000	0.000	0.000	0.000	0.000	0.000	0.000	0.000	0.000	0.000	0.000	0.000	0.000	0.000
H2	0.526	0.332	1.000	0.424	0.000	0.000	0.000	0.000	0.000	0.000	0.316	0.616	0.301	0.000	0.000	0.000	0.000	0.000	0.000	0.000
H2S	0.000	0.000	0.000	0.000	0.000	0.000	0.000	0.000	0.000	0.000	0.000	0.000	0.000	0.000	0.000	0.000	0.000	0.000	0.000	0.000
S8	0.000	0.000	0.000	0.000	0.000	0.000	0.000	0.000	0.000	0.000	0.000	0.000	0.000	0.000	0.000	0.000	0.000	0.000	0.000	0.000
C2H4	0.000	0.002	0.000	0.002	0.000	0.000	0.000	0.000	0.000	0.000	0.003	0.003	0.003	0.000	0.000	0.000	0.000	0.000	0.000	0.000
C2H6	0.000	0.001	0.000	0.001	0.000	0.000	0.045	0.230	0.000	0.000	0.001	0.001	0.006	0.000	0.000	0.000	0.000	0.000	0.000	0.000
C3H6	0.000	0.001	0.000	0.002	0.000	0.000	0.000	0.000	0.000	0.000	0.003	0.003	0.003	0.000	0.000	0.000	0.000	0.000	0.000	0.000
C3H8	0.000	0.000	0.000	0.001	0.000	0.000	0.046	0.230	0.000	0.000	0.001	0.001	0.006	0.000	0.000	0.000	0.000	0.000	0.000	0.000
ISOBU-01	0.000	0.000	0.000	0.000	0.000	0.000	0.023	0.115	0.000	0.000	0.000	0.000	0.003	0.000	0.000	0.000	0.000	0.000	0.000	0.000
N-BUT-01	0.000	0.000	0.000	0.000	0.000	0.000	0.023	0.115	0.000	0.000	0.001	0.001	0.003	0.000	0.000	0.000	0.000	0.000	0.000	0.000
1-BUT-01	0.000	0.001	0.000	0.002	0.000	0.000	0.000	0.000	0.000	0.000	0.002	0.002	0.002	0.000	0.000	0.000	0.000	0.000	0.000	0.000
1-PEN-01	0.000	0.001	0.000	0.000	0.000	0.000	0.000	0.000	0.045	0.000	0.000	0.000	0.000	0.000	0.000	0.000	0.145	0.000	0.000	0.000
N-PEN-01	0.000	0.000	0.000	0.000	0.000	0.000	0.046	0.000	0.105	0.000	0.000	0.000	0.000	0.000	0.000	0.000	0.016	0.000	0.000	0.000
1-HEX-01	0.000	0.001	0.000	0.000	0.000	0.000	0.000	0.000	0.039	0.000	0.000	0.000	0.000	0.000	0.000	0.000	0.124	0.000	0.000	0.000
N-HEX-01	0.000	0.000	0.000	0.000	0.000	0.000	0.048	0.000	0.105	0.000	0.000	0.000	0.000	0.000	0.000	0.000	0.013	0.000	0.000	0.000
C7-C11	0.000	0.002	0.000	0.000	0.000	0.000	0.254	0.000	0.706	0.000	0.000	0.000	0.000	0.000	0.000	0.000	0.472	0.000	0.000	0.000
C12-C18	0.000	0.001	0.000	0.000	0.000	0.000	0.346	0.000	0.000	0.772	0.000	0.000	0.000	0.000	0.000	0.000	0.230	0.000	0.000	0.000
C19-C24	0.000	0.001	0.000	0.000	0.000	0.066	0.112	0.000	0.000	0.228	0.000	0.000	0.000	0.000	0.000	0.000	0.000	0.000	0.000	0.000
C25PL	0.000	0.010	0.000	0.000	0.000	0.934	0.000	0.000	0.000	0.000	0.000	0.000	0.000	0.000	0.000	0.000	0.000	0.000	0.000	0.000
ALCS	0.000	0.001	0.000	0.000	0.004	0.000	0.000	0.000	0.000	0.000	0.000	0.000	0.000	0.000	0.000	0.000	0.000	0.000	0.000	0.000
Solids Mass Flow lb/hr	249,961	249,961	668	162,341	33,397	48,347	49,015	2,342	15,233	36,915	3,386	946	36,013	318,262	441	489,676	6,176	626,335	99,500	156,301
Enthalpy MMBtu/hr	-377.6	-672.7	0.1	-464.4	-226.7	-39.6	-36.8	-2.8	-14.6	-32.6	-6.2	-1.2	-46.8	-1242.0	-3.0	8.8	-4.9	-350.0	-391.9	-863.4
Pressure psi	295.0	295.0	330.0	295.0	295.0	295.0	295.0	295.0	14.7	14.7		295.0	295.0	2200.0	23.0	16.1	295.0	16.1	2200.0	500.0
Temperature F																				
Pressure psi	295.0	295.0	330.0	295.0	295.0	295.0	295.0	295.0	14.7	14.7		295.0	295.0	2200.0	23.0	16.1	295.0	16.1	2200.0	500.0
Mass Flow lb/hr	0	0	0	0	0	0	0	0	0	0	0	0	0	0	0	0	0	0	0	0
Enthalpy MMBtu/hr																				

End Notes

[1] The 8wt% biomass in the feed stream accounts for 5% of the total feedstock energy (higher heating value (HHV) basis).

[2] Transportation fuels are responsible for most GHG emissions by end-use sector. By comparison, electric power generation is responsible for more total CO_2 emissions (2,433 versus 2,014 million metric tons per year in 2007), but electricity is used across many different end-use sectors. [1,2]

[3] This accounting recognizes the net-negative CO_2 emissions associated with sequestering photosynthetic CO_2 derived from biomass. Carbon credits are not taken from the buildup of carbon in the soil and roots of the biomass as a widely accepted accounting method for this carbon has yet to be finalized.

[4] This is achieved at a thirty percent (by weight) biomass feed, which has been demonstrated at the commercial scale in Buggenum, The Netherlands, and the application of aggressive CCS practices [4].

[5] The Required Selling Price (RSP) is the minimum price fuels can be sold for in order to: a) offset operating costs, b) service its debt, and c) provide the expected rate of return to investors. See Section 4.2 for an in-depth discussion on this topic.

[6] The 8wt% and 15wt% biomass cases equate to 5% and 10% of the feed streams on an HHV basis.

[7] SG refers to switchgrass, the biomass type used in this study.

[8] Commercial indirect liquefaction processes have largely used coal as a feedstock due to its low cost, high energy density, and ease of availability. However, the process can use any feedstock which can be gasified, including biomass.

[9] As opposed to a dilute stream of CO_2 as produced by many other energy conversion and fuel production systems.

[10] This incremental cost includes the cost of CO_2 compression, transport, sequestration, and 80 years of monitoring.

[11] An increased interest in reducing GHG emissions has resulted in the addition to the plant shown in red. This addition – the "CCS+ATR" configuration – uses existing technologies already found in commercial CTL plants to enable increased CO_2 capture over previous plant designs.

[12] The advantage of having separate feed systems would be that, if the biomass system becomes inoperable for a time because of plugging, the gasifier can still continue to operate on coal only.

[13] The FT reactors require a 1:1.0 to 1.1:1 ratio of H_2 to CO in the syngas for proper operation. In the CTL and CBTL cases, the syngas exiting the gasifier has a 0.4 H_2:CO and therefore requires the water gas shift in order to ensure proper operation.

[14] The product mix from a FT reactor varies widely with operating conditions. Specifically, higher temperature operation produces shorter chain hydrocarbons and a larger number of oxygenates, a product mix which is better suited for gasoline production.

[15] Light hydrocarbons cannot be captured in the downstream MDEA unit. The partial oxidation, i.e. partial combustion, of light hydrocarbons allows the recovery of their thermal heating value while converting them into a form which is useful in the recycle stream (CO). Furthermore, if that CO is shifted to CO_2 using a WGS unit, it can be captured by the MDEA unit.

[16] The field is expected to produce 155 million barrels of incremental oil over 30 years of CO_2 flooding, or an average of 5 million bbl/yr of incremental oil, and the project is in its eighth year. Assuming that incremental production did not occur until year two of the project, an estimated 30 million barrels of incremental oil have been produced. This ramp-up seems commiserate with a current incremental production rate of 18,000 bbl/d, or 6 million bbl/yr [12].

[17] The production of diesel and jet fuel is of particular importance given the lack of alternatives for certain vehicles which use these fuels. In particular: aircraft, long haul trucks, and trains cannot be easily retrofitted to use hybrid electric technologies or configured for different fuel usage.

[18] CMM reduction methods are currently in practice at a number of domestic mines and can be used to produce saleable gas. In many cases, however, low grade CMM is used for mine energy requirements, including the heating of ventilation air [14].

[19] Title II, Subtitle A, Sec. 201 of EISA 2007 states that alternative transportation fuels are to be compared to the "average life cycle GHG emissions, as determined by the [EPA] Administrator, after notice and opportunity for comment, for gasoline or diesel fuel (whichever is being replaced by the renewable fuel) sold or distributed as transportation fuel in 2005."

[20] The "well-to-tank" emissions are the emissions associated with the production and delivery of diesel fuel.

[21] Vehicle operation is included in the petroleum baseline in accordance with the EISA 2007 definition of "baseline life cycle greenhouse gas emissions."

[22] In reality, it is not possible to eliminate all upstream emissions.

[23] Depending on the type of biomass, the cultivation methodology, and the soil quality, some of the plant carbon may be deposited in the soil and/or left in the soil as roots. Although this phenomenon has been detailed in the literature, GHG credits are not taken for soil and root carbon in this study. This is due to a lack of consensus in the field as to the proper accounting methodology for how much CO_2 is converted and remains in the ground. [17].

[24] The co-gasification of 30% (by weight) biomass is the largest biomass percentage which has been demonstrated for long operating times in a commercial-scale gasifier.

[25] See Appendix B for more information on the disposition of carbon exiting the plant in the naphtha.

[26] Numerous opportunities exist for increasing the carbon conversion in the CTL/CBTL/BTL process, thereby reducing the fraction of carbon converted into CO_2. These include the use of green energy sources to power process equipment and the integration of natural or green processes which generate hydrogen or oxygen. By integrating these processes into the CTL or CBTL plant, less coal is required to power process equipment, resulting in higher conversion percentages. Furthermore, a majority of the research into improving Integrated Gasification Combined Cycle (IGCC) clean coal power plants is applicable to CTL/CBTL/BTL and can result in similar improvements in conversion rates, which would also reduce the GHG emissions profile of the fuel.

[27] This includes the cost of CO_2 compression, transport, sequestration, and monitoring of the CO2 for 80 years.

[28] The RSP is the product price at the plant gate and does not include transportation costs or taxes.

[29] The following costs are included in the total overnight capital cost: bare erected cost, engineering and construction management fees, process and project contingencies, inventory capital, preproduction (startup) costs, and other miscellaneous owner's costs.

[30] As mentioned in Chapter 2, CO_2 is captured part of the CBTL process. In the non-CCS cases, however, this captured CO_2 is vented.

[31] The capacity is limited to 30,000 bpd due to the 4,000 dry tons per day biomass feed rate limit detailed in Chapter 2.

[32] Some standards may allow fuels that significantly exceed the GHG emission reduction standard to be blended with non-qualifying fuels to yield a fuel mixture that does qualify.

[33] California's LCFS requires fuels to achieve a reduction of at least 10% by 2020 and "more thereafter". This LCFS is contained in the Global Warming Solutions Act of 2006, California Assembly Bill 32 [18, 19].

[34] In May 2007, Senator Barack Obama and Senator Tom Harkin introduced legislation calling for a National Low Carbon Fuel Standard that would require fuel refiners to reduce life-cycle greenhouse gas emissions of the transportation fuels sold in the U.S. by five percent in 2015 and ten percent in 2020 [20]. During his presidential campaign, Senator Obama reiterated his call to establish a National LCFS in the "Obama-Biden New Energy Plan for America" [21].

[35] The EISA 2007 Renewable Fuels Standard requires qualifying bio-fuels to have life-cycle GHG emissions that are 20% below the petroleum baseline.

[36] In the case of any legislation involving a LCFS, it is important to define what baseline is being used, including the base year and what methodology is to be used in defining that baseline. As shown in Chapter 3, baselines vary widely and can influence the "Go" vs. "No Go" tipping point for a technology.

[37] Section 526 states "No Federal agency shall enter into a contract for procurement of an alternative or synthetic fuel, including a fuel produced from nonconventional petroleum sources, for any mobility-related use, other than for research or testing, unless the contract specifies that the lifecycle greenhouse gas emissions associated with the production and combustion of the fuel supplied under the contract must, on an ongoing basis, be less than or equal to such emissions from the equivalent conventional fuel produced from conventional petroleum sources."

[38] See Figures 4-6 and 4-7 above for more information.

[39] The Renewable Fuels Standard under EISA 2007 uses a 2005 petroleum baseline for the basis of comparison, identical to that used in this study.

[40] The significant contribution in GHG emissions reductions from CTL+CCS in the Section 526 Requirement scenario are the result of both the 5% emissions reduction compared to the petroleum baseline and the large amount of FT fuel produced (191 million bpy). Even though the emissions reduction is higher in the BTL+CCS case (322% below petroleum), only 9 million bpy of diesel fuel is produced, limiting the total emissions reductions from BTL.

[41] An anonymous reviewer provided this insight.

[42] "Net surplus" is the difference between the change in consumer surplus less the change in producer surplus. A decline in world oil prices benefits consumers and hurts domestic oil producers, but on balance this effect is positive for the country.

[43] Unlike the EIA, the IEA does not present a "high oil price case."

[44] The number of plants can be deduced by dividing the production rate by the notional plant size of 50,000 b/d. Thus over the five year period between 2015 and 2020, sixteen, twenty-four, and forty 50,000 b/d plants would have to be added under the 2mm, 3mm, and 5mmb/d ramp-ups, respectively.

[45] See Southern States Energy Board's *American Energy Security,* and RAND, 2008, for a discussion of the issues [27, 28].

[46] *Economic Impacts of U.S. Liquid Fuel Mitigation Options*, p. 45, estimates 350,000 jobs for production of 2 million bpd and 500,000 jobs at 5 million bpd [29]. Similarly, in an analogous report, RAND Corporation estimates 200,000-300,000 jobs in a 3 million bpd shale oil industry (*Oil Shale Development in the United States*, pp. 27-28) [30]. This stands in contradistinction to the "no net effect" in RAND's 2008 CTL report [27]. Finally, the *American Energy Security*, p. 151, estimates over 900,000 jobs from 8.4 million bpd of alternative fuels, inclusive of coal to liquids, shale oil, and biomass to liquids [28].

[47] Financial incentives such as tax incentives, loan guarantees, and other mechanisms may also play a role in addressing the economic and market challenges of CBTL, but these are beyond the scope of this paper.

[48] Gasification using high-temperature and high-pressure entrained flow gasifiers should help to eliminate tar and methane formation from the biomass component of the feedstock. Also, the CBTL plants would be simpler and less costly if the same gasifier could be used to process both the coal and the biomass.

[49] The CTL with CCS plant configuration produces fuel which has 5% less life cycle GHG emissions than petroleum-derived diesel, whereas 15wt% CBTL with CCS produces fuel with emissions that are 33% less than petroleum-derived diesel.

[50] EIA, Annual Energy Outlook 2009 (early release), December 2008, reference case

[51] As-received switchgrass is assumed to have a moisture content of 10% by weight, so the as-received cost can be calculated by multiplying the dry cost by a factor of 0.9.

[52] This is an approximation. Diesel retail prices, and to a lesser extent the differences in ULSD and LSD retail prices, vary because of state taxes.

[53] ISO 14040 LCA standard recommends "system expansion" as a preferred method to avoid allocation of emission to co-products. The displacement (or substation) method is a form of system expansion.

[54] The amount of methane that remains in the coal is primarily a function of the amount of pressure the coal seam is under, and hence the depth of that seam. The deeper the coal seam, the greater the pressure and the more methane that remains physically absorbed to the coal. In shallow seams, the methane naturally migrates out of the coal seam and eventually into the atmosphere. Migration occurs in deeper seams, but the rate is limited such that migration is limited to the surrounding rock strata, and methane content remains high in the coal.

[55] Best practice CMM recovery methods include the drilling of gob wells (used by more than 21 mines in the U.S.) and longhole horizontal boreholes (used by over 10 U.S. mines) [EPA 1999]

[56] Ortiz, D. S.; Willis, H. H.; Pathak, A.; Sama, P.; Bartis, J.T.; *Characterization of Biomass Feedstocks*, DRR-4440- NETL, RAND Corporation, 2008

[57] As will be seen, the Variable Operating Costs are dominated by fuel costs, and Variable O&M dominates the total O&M costs. Hence efficiency can have a significant effect on operating costs.

[58] A portion of the gas exiting the FT reactor is combusted to create power for the plant and the remainder is recycled back to the FT reactor for conversion of the carbon into liquid transportation fuels. However, a maximum of 76% of the tail gas exiting the reactor can be recycled back to the reactor. Any tail gas in excess of this is combusted to prevent the build-up of gases in the process.

[59] Process efficiency is not related to plant scale at the plant scales considered.

In: Clean Energy Solutions from Coal
Editor: Marcus A. Pelt
ISBN: 978-1-61324-724-2

Chapter 2

HYDROGEN FROM COAL PROGRAM*

United States Department of Energy

EXECUTIVE SUMMARY

Hydrogen has the potential to play a significant role in the nation's energy future, particularly for the production of clean electric power from coal. Production and use of hydrogen in a gasification combined cycle system for stationary power applications will complement the development of next generation hydrogen turbine technology that enables the plant to achieve near-zero pollutant emissions and increased plant efficiency. The Hydrogen from Coal Program's Research, Development, and Demonstration (RD&D) activities include development of hydrogen separation membranes and other advanced technologies that efficiently produce high purity hydrogen for stationary power production. When combined with carbon management technologies such as carbon capture and storage (CCS) and coal-biomass co-utilization, these next generation power plants will achieve significant reductions in greenhouse gas (GHG) emissions with low electricity costs.

THE HYDROGEN FROM COAL RD&D PLAN

This multi-year RD&D Plan addresses the strategies, goals, milestones and progress of the program, and defines the research areas needed to support the other Office of Clean Coal (OCC) programs and the overall DOE Hydrogen activity.

This RD&D Plan is organized by section, as follows:

Section 1. Introduction
Section 2. Hydrogen from Coal Program Mission and Goals
Section 3. Technical Discussion Section 4. Technical Plan

---

* This is an edited, reformatted and augmented version of a United States Department of Energy publication, Research, Development and Demonstration Plan for the period 2010 through 2016, External Draft, dated September 2010.

Section 5. Implementation Plan

Detailed activities and technical targets are provided in the Technical Plan in Section 4. Implementation of the Program's activities will be coordinated closely with the related activities supported by the Office of Fossil Energy (FE) and other organizations both inside and outside the government.

The FE Hydrogen from Coal Program was initiated in fiscal year 2004 (FY 2004). The Program is transitioning from hydrogen production for transportation applications to electric power applications by reducing technological market barriers for the reliable, efficient, and environmentally friendly conversion of coal to hydrogen with carbon capture and storage and other carbon reduction techniques. This Hydrogen from Coal RD&D Plan focuses on those hydrogen activities necessary to support the goals of FE's Office of Clean Coal in development and demonstration of advanced, near-zero emission coal-based power plants. The outcome of this strategy will be the deployment of advanced hydrogen separation membrane modules and other advanced concepts such as process intensification that provide high purity hydrogen for use in a stationary turbine using integrated gasification combined cycle (IGCC) technology. Collectively, these technologies will enhance plant efficiency and in combination with CCS and displacement of a portion of the coal with biomass, provide significantly reduced GHG emissions.

Goals: The goals of the Hydrogen from Coal Program RD&D activities are:

- Prove the feasibility of a 40 percent efficient, near-zero emissions power facility that uses membrane separation technology as well as other advanced technologies to reduce the cost of electricity by at least 35 percent (relative to a base case IGCC with CCS using currently available technologies).
- Develop hydrogen production and processing technologies that will contribute approximately 2.9 percent in improved efficiency and 12 percent reduction in cost of electricity to the 40 percent efficient near-zero emissions power facility.

Overview of Technology in the RD&D Plan

RD&D will be directed toward hydrogen production and separation from coal gasification for stationary power applications by:

- Performing research on:
    - Hydrogen membrane separation, and
    - New strategies for process intensification and other advanced concepts such as coproduction and chemical looping.

Advanced hydrogen membranes integrated with other advanced IGCC technologies can improve overall power generation efficiency and reduce costs. Process intensification involves developing novel technologies that combine multiple processes into one step, use new control methods, or integrate alternative energy technologies with hydrogen from coal technologies. Chemical looping offers the potential for hydrogen production with near-100 percent carbon capture and increased plant efficiency. Novel co-production concepts may allow more effective plant operation or further reduce costs for clean electric power production.

- Pursuing technical targets for the program including the following:
  - o Develop hydrogen membranes by 2015 that can provide hydrogen fuel that meets gas turbine specifications from syngas under warm gas cleanup conditions at a flux level of 300 standard cubic feet per hour per square foot ($SCFH/ft^2$) of membrane area, at pressures that will reduce compression costs for H_2 and CO_2, and at a membrane cost of less than $100 per ft^2.
  - o Develop advanced concepts for syngas processing in gasification-based electric power generation systems that will significantly reduce complexity and cost, improve plant and fuel utilization and/or improve efficiency, to produce H_2 for gas turbine fuel and CO_2 for sequestration.

Accomplishments and Progress

The Hydrogen from Coal Program has successfully transitioned from its initial start-up in FY 2004 to full operations. The Program has been actively soliciting proposals from industry, universities, and other organizations to help the program achieve its goals. Currently, the program has projects to develop advanced technologies targeted toward higher efficiency and reduced cost of electricity from IGCC plants that produce hydrogen and use it to generate electricity.

Research progress is periodically reviewed to update the RD&D Plan with respect to goals, technical targets, milestones, and program schedules. This FY 2010 update reflects the focus on close coordination with the Office of Clean Coal power generation programs and activities to achieve lower cost, improved plant efficiency, reduced GHG emissions, and creation of jobs.

Technical

Ongoing research activities are aimed at development and scale-up of precious-metal-based hydrogen separation membranes and other membrane concepts. Several of the hydrogen membrane developers' test results have shown that, in the absence of sulfur, their membranes can exceed the 2015 flux technical target noted above. Notably:

- Eltron Research, Inc. is developing alloy-based membranes and has developed a separator unit rated to produce 1.5 lbs/day of hydrogen. Eltron's best alloy membrane has demonstrated a H_2 flux rate of 411 $SCFH/ft^2$ at specified pressure and gas compositions.
- Worcester Polytechnic Institute (WPI) is developing Pd-based membranes on tubular stainless steel or alloy supports. WPI achieved a H_2 flux of 359 $SCFH/ft^2$ in pure gases at 442°C and 100 psi ΔP with a 3–5 micron (µm) thick palladium(Pd) membrane with an Inconel base. WPI has also built an engineering-scale prototype membrane.
- Praxair is building a Pd-alloy based prototype multi-tube hydrogen purifier and will use the unit to demonstrate prototype performance. A Pd-Au membrane (5 percent

Au, 9μm thickness) showed H_2 flux in pure gas of 384 SCFH/ft^2 and the H_2/N_2 selectivity was 495 at 200 psi.

- United Technologies Research Center (UTRC) is developing two types of membrane separators using metallic supports: one based on a commercial tube design with a Pd-based alloy; the other using a novel nano-composite material. UTRC's current membrane flux performance is approximately 45 SCFH/ft^2 at a temperature of 450°C and feed pressure of 200 psi. Anticipated performance for the nano-oxide membrane is H_2 flux of 400 SCFH/ft^2.
- Southwest Research Institute (SwRI) is developing Pd-foil-based membrane separators. SwRI has tested 70 percent Pd, 20 percent Au, 10 percent Pt foils under the DOE Test Protocol gas composition, with flux levels of approximately 30 SCFH/ft^2.

Precious metals are effective materials for hydrogen separation membranes. However, they are expensive and present a potential security risk, given that some supplies are imported. The National Energy Technology Laboratory (NETL) completed a comprehensive assessment of the production of precious metals currently used in hydrogen membrane fabrication. The assessment showed:

- Commercial deployment using precious metals has potential global economic and environmental impacts.
- Global deposits are primarily located in South Africa and Russia.
- Less than 10 significant mining companies currently exist in the world; production is declining.

Collectively, these factors could restrain the ability to provide for affordable and environmentally acceptable hydrogen production via membrane separation technologies. To address these concerns, a competitive Funding Opportunity Announcement (FOA) was released soliciting research projects that would conduct both fundamental and applied research on novel, non-precious metal hydrogen separation technologies. The projects were selected in September 2009 and are shown in Table ES-1.

Based on its earlier research work, the Hydrogen from Coal Program issued a FOA in March 2010 seeking innovative projects that could be scaled up for eventual pre-commercial demonstration. The FOA was closed in May 2010 and four projects were selected: Praxair, Inc.; United Technologies Research Center; Western Research Institute; and Worcester Polytechnic Institute.

Activity

Hydrogen production from coal activities are closely linked with the system's up-front gasification technologies, downstream turbine combustion, and CO_2 capture and sequestration. Therefore, these four DOE programs are coordinated within the Office of Clean Coal to enhance integration of the separate programs. Additionally, the Hydrogen from Coal Program continues to coordinate with other DOE offices by participating in the development of various planning documents and in the DOE Hydrogen Program and Vehicle Technologies Program Annual Merit Review of the sponsored projects.

Table ES-1. Non-precious Metal Membrane Projects

Project Performer	Project Title/Description	DOE Funding	Participant Funding
Colorado School of Mines	Nanoporous metal carbide surface diffusion membranes	$998,543	$250,628
Ceramatec	Ceramic or ceramic composite proton conducting membranes	$924,549	$253,637
Worcester Polytechnic Institute	Supported molten metal membranes	$996,567	$249,857
Southwest Research Institute	Amorphous Zr-Ni alloy membranes for hydrogen separations	$799,786	$199,950
University of Florida	Novel magnetically assisted fluidized bed reactor development for chemical looping	$999,920	$249,980
University of Nevada-Reno	Amorphous alloy membranes prepared by melt-spin methods	$1,163,596	$290,899
University of Texas at Dallas	Non-precious metal mixed matrix membranes	$1,000,000	$250,000

Benefits

The technologies being developed by the Hydrogen from Coal Program offer a variety of important technical, economic, and environmental benefits. These benefits include a reduced carbon footprint, reduced cost of electricity, increased energy security through reduced imports, and the creation of high-tech domestic jobs.

Reduced Carbon Footprint

Gasification technologies have shown the potential to produce clean hydrogen and electricity from coal and coal/biomass mixtures with virtually zero criteria pollutant emissions. Hydrogen turbines utilized in combined cycle power production systems using hydrogen from gasification can provide electricity at higher efficiency and lower cost than conventional systems, while facilitating the capture of carbon dioxide for storage. Carbon sequestration technologies are being developed to provide the capability to cost-effectively use the concentrated CO_2 streams from gasification in enhanced oil recovery, geological storage, and accelerated biomass growth processes for fuel production[1].

Substitution of a portion of the coal feedstock with biomass can provide an added benefit in carbon reduction. NETL systems studies show that the combination of CCS and substitution of a portion of the coal feedstock with biomass of several types will provide substantially lower life-cycle carbon emissions, while allowing operators to use regional non-food biomass resources in large efficient coal-based plants.

Reduced Cost of Electricity

When compared to the reference IGCC case with CCS using current technology, an advanced IGCC plant with CCS using warm gas cleanup, hydrogen separation membranes and an advanced hydrogen turbine offers a significant improvement in plant efficiency and capital costs, leading to reduced cost of electricity to consumers. The membranes are less expensive than the technology they replace, and provide the added benefit of producing CO_2 at a higher pressure, leading to reduced compression costs for sequestration. Overall, NETL studies estimate that use of hydrogen separation membranes in conjunction with warm gas cleanup instead of the current Selexol technology for separation of CO_2 and H_2 will improve plant efficiency by 2.9 percent and reduce the overall cost of electricity by 12 percent.[2]

Co-production of electricity with hydrogen, other fuels, or chemicals for export could also reduce electricity costs. Hydrogen could be exported and sold to reduce the cost of electric power, improve gasifier utilization, and help build hydrogen infrastructure. Production and storage of hydrogen at the plant site during periods when electricity is not needed (e.g., night) would allow higher capacity electric power production when electricity demand is high (e.g., peak daytime hours) or allow better integration with intermittent sources such as solar or wind.

Creation of High-Tech Domestic Jobs

The development of hydrogen production and separation technologies that provide high purity hydrogen will complement the development of advanced hydrogen turbines. This will result in the United States becoming a key leader in these technologies and creation of new high paying domestic jobs to manufacture and oversee the deployment and operation of next generation gasification and turbines technology.

Increased Energy Security

The technologies developed by the Hydrogen from Coal Program offer the potential to improve domestic energy security in two ways. The electrification of the nation's energy sectors using clean, highly efficient, low carbon coal power production systems to supply surface transportation vehicles (e.g., plug-in hybrids) will over time reduce the need for insecure imported crude oil. Additionally, while offering reduced cost of electricity benefits noted above, co-production technology provides the potential to domestically produce a variety of fuel and chemical products which are currently imported.

Technical Activity Gantt Chart Summary

The specific sub-element activities and their associated timelines are shown in the chart in Figure ES-1, which summarizes the activities and technologies associated with hydrogen production from large-scale IGCC plants.

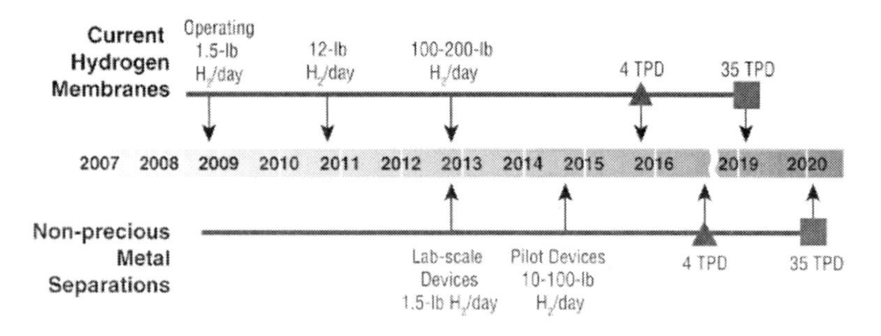

Figure ES-1. Hydrogen from Coal RD&D Program Milestones.

Current Hydrogen Membranes

End of 2008: Operating at 1.5 lb H_2/day

End of 2010: Operating at 12 lb H_2/day

End of 2012: Operating at 100–200 lb H_2/day

End of 2015: Scale-up membrane module to 4 tons H_2/day production for component testing Beginning of 2019: Scale-up membrane module to 35 tons H_2/day production for integrated testing

Non-precious Metal Separations

End of 2012: Operating lab-scale devices at 1.5 lb H_2/day

End of 2014: Operating pilot-scale devices at 10–100 lb H_2/day

End of 2016: Scale-up non-precious metal membrane module to 4 tons H_2/day for component testing Beginning of 2020: Scale-up non-precious metal membrane module to 35 tons H_2/day for integrated testing

1. Introduction

As a preeminent primary source of energy, coal is an abundant domestic resource, with the United States boasting hundreds of years of supply at current demand levels. Today coal represents over 50 percent of the nation's electricity supply, but contributes over 40 percent to the nation's total carbon dioxide emissions. Continued use of coal-supplied electricity will require the significant reduction of the carbon footprint of coal-fired power plants. As seen in Figure 1, the gasification of coal to produce hydrogen and its use in turbines in an IGCC system provides a pathway to produce clean electricity from coal with higher efficiency, lower carbon footprint and lower cost when compared to other coal power generation technologies. In combination with co-feeding of biomass and carbon management techniques including carbon capture and sequestration and advanced re-use such as algae production, these technologies could provide electric power with near-zero emissions.

The Hydrogen from Coal Program RD&D Plan provides a roadmap that the program will pursue to develop the technologies necessary for coal to meet the overall OCC goals of reduced GHG emissions, favorable economics, new clean energy jobs, and improved energy security. The advanced power generation program goals for OCC are to achieve 90 percent carbon capture while maintaining less than 10 percent increase in cost of electricity (COE) over a 2003 reference IGCC plant having no carbon capture. The COE of that plant is 9.3¢/kWh, so the cost target for carbon capture is no more than 10 percent greater, or 10.2¢/kWh. The IGCC case in Figure 1 represents technology progress as of 2007. Integration of technologies being researched under the Hydrogen from Coal Program will play a key role in achieving the cost reduction goal.

The Plan discusses current and future technologies for the production of hydrogen from coal, its use for producing clean electricity, and its potential for the co-production of hydrogen and power with near-zero emissions. The Plan will serve as a resource document for the hydrogen activities, emphasizing close coordination with OCC goals, milestones, and targets.

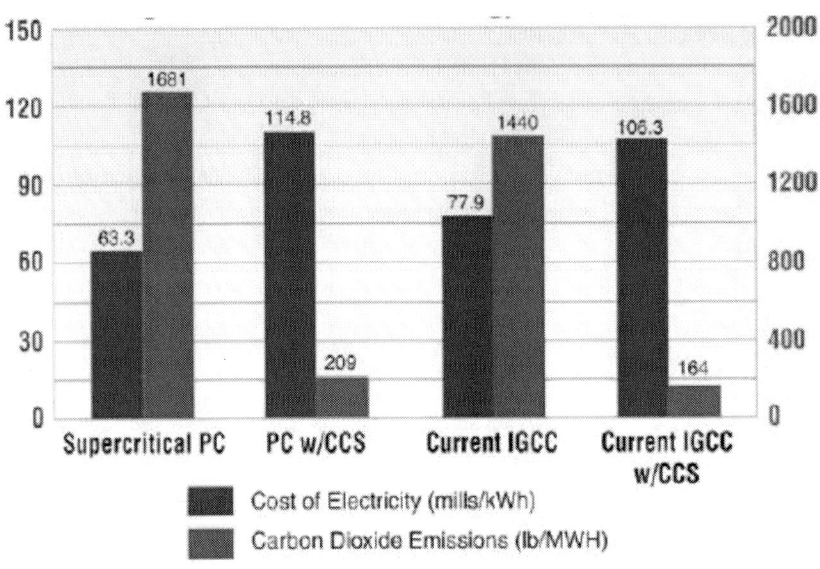

Source: DOE/NETL, Cost and Performance Baseline for Fossil Energy Plants, 2007
Note: 1 mill = 1/10th of one cent.

Figure 1. Cost of Electricity and Environmental Footprint of Current Technology Coal Plants.

2. Hydrogen from Coal Program – Mission and Goals

The mission of the Hydrogen from Coal Program is to develop advanced and novel hydrogen production technologies that will ensure the use of our nation's abundant coal and biomass resources to produce affordable electricity in a safe and environmentally clean manner. The RD&D activities will provide the pathways to produce affordable hydrogen from coal in an environmentally clean manner. These technologies can provide reduced carbon footprint and lower cost of electricity. When compared to stand-alone electricity, fuels, or chemical facilities, the coproduction of electricity and hydrogen from coal or coal/biomass mixtures offers a variety of important technical, economic, and environmental benefits.

The goals for the Hydrogen from Coal Program are:

- Prove the feasibility of a 40 percent efficient, near-zero emissions power facility that uses membrane separation technology as well as other advanced technologies to reduce the cost of electricity by at least 35 percent (relative to a base case IGCC with CCS using currently available technologies).
- Develop hydrogen production and processing technologies that will contribute approximately 2.9 percent in improved efficiency and 12 percent reduction in cost of electricity for the 40 percent efficient near-zero emissions power facility.

3. Technical Discussion

3.1. Current Gasification Technology

Hydrogen can be produced from coal by gasification followed by processing the resulting synthesis gas using currently available technologies. The coal is first gasified with oxygen and steam to produce a synthesis gas consisting mainly of carbon monoxide (CO) and hydrogen (H_2), with some CO_2, sulfur, particulates, and trace elements. Oxygen (O_2) is added in less than stoichiometric quantities so that complete combustion does not occur. This process is highly exothermic, with temperatures controlled by the addition of steam. Increasing the temperature in the gasifier initiates devolatilization and breaking of weaker chemical bonds to yield tars, oils, phenols, and hydrocarbon gases. These products generally further react to form H_2, CO, and CO_2. The fixed carbon that remains after devolatilization is gasified through reactions with O_2, steam, and CO_2 to form additional amounts of H_2 and CO. These gasification reactions are shown in Figure 2.

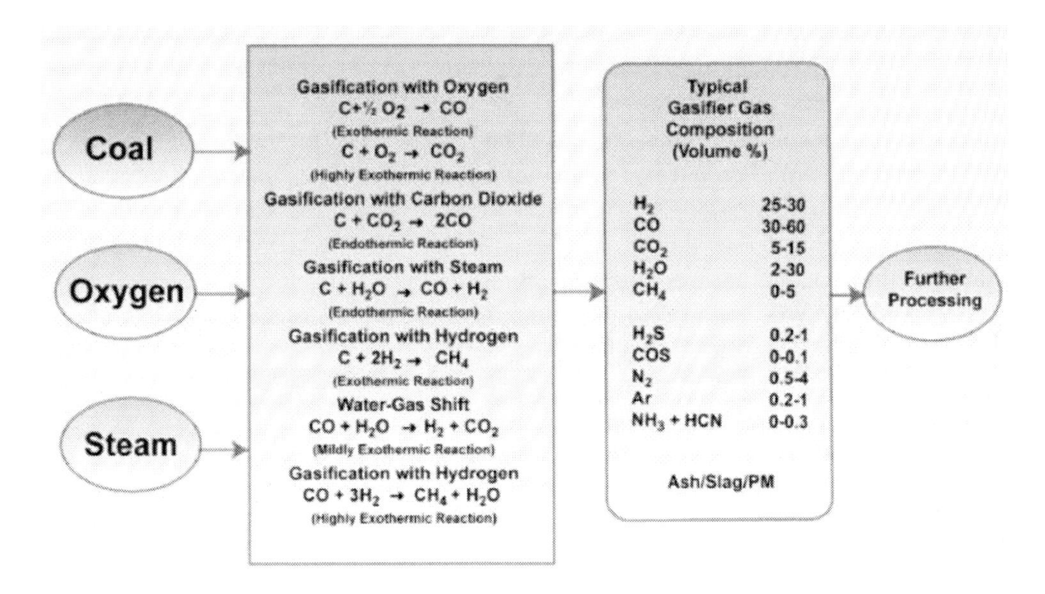

Figure 2. Major Gasification Reactions.

The minor and trace components of coal also are transformed in the gasification reactor. Under the substoichiometric reducing conditions of gasification, most of the fuel's sulfur converts to hydrogen sulfide (H_2S), but some (3–10 percent) also converts to carbonyl sulfide (COS). Nitrogen bound with the fuel generally converts to gaseous nitrogen (N_2), with some ammonia (NH_3) and a small amount of hydrogen cyanide (HCN) also being formed. Most of the chlorine content of the fuel is converted to hydrogen chloride (HCl) gas and some particulate-phase chlorides.

Minerals in the feedstock (ash) separate and leave the bottom of the gasifier as an inert slag (or bottom ash), a potentially marketable solid product. The fraction of the ash entrained with the syngas, which is dependent upon the type of gasifier employed, requires removal downstream in particulate control equipment, such as filters and water scrubbers. This particulate is typically recycled to the gasifier to ensure high carbon conversion. Some gasifiers also yield devolatilization or pyrolysis products (e.g., coal tars, oils, phenols), some of which can be sold. The remaining products can and must be controlled to eliminate any potential environmental impacts.

Trace elements associated with both organic and inorganic components of the coal, such as mercury and arsenic, are released during gasification and settle in different ash fractions (e.g., fly ash, bottom ash, slag) and gaseous emissions. The particular chemical species and physical forms of condensed-phase and vapor-phase trace elements are functions of gasifier design and operating conditions.

The temperature of the synthesis gas as it leaves the gasifier is generally in the range of 1,000°F to 1,900°F, depending upon the type of gasifier selected. With current technology, the gas has to be cooled to ambient temperatures to remove contaminants, although with some designs, steam is generated as the synthesis gas is cooled. Depending on the system design, a scrubbing process is used to remove HCN, NH_3, HCl, H_2S, and particulates, and operates at low temperatures with synthesis gas leaving the process at about 72°F. The H_2S and COS, once hydrolyzed, are removed by dissolution in, or reaction with, an organic solvent and converted to valuable by-products, such as elemental sulfur or sulfuric acid with 99.8 percent sulfur recovery. The residual gas from this separation can be combusted to satisfy process-heating requirements.

The raw clean synthesis gas must be re-heated to 600–700°F for the first of two water-gas shift (WGS) reactors that produce additional hydrogen from water through the catalytically assisted equilibrium reaction of CO with H_2O to form CO_2 and H_2. The exothermic reaction in the WGS reactor increases the temperature to about 800°F, which must be cooled to the required inlet temperature for the second WGS reactor in the range of 250–650°F, depending on design. The WGS reaction increases the amount of product H_2 in the final mixture, as well as the concentration of carbon in a single product (CO_2), which allows for easier capture. Hydrogen must be separated from the shifted gas containing CO_2, CO, and other contaminants, and may need to undergo a polishing step that removes any remaining sulfur, CO, and other trace contaminants in order to meet the requirements for various end-uses (e.g., turbines or fuel cells). The resulting hydrogen can provide excellent fuel for advanced gas turbines with no carbon emissions, while the captured CO_2 can be routed to compression and sent to sequestration facilities. The specifications for fuel gas for advanced gas turbines are shown in Table 1. The available information shows that there is the need for more data to determine the purity requirements for hydrogen separation membranes for gas turbines and other relevant applications.

Hydrogen could also be exported to improve gasifier utilization and potentially reduce the cost of electric power, depending upon product prices. Production and storage of hydrogen at the plant site during periods when electricity demand is low would allow higher capacity electric power production when electricity demand is high (e.g., peak daytime hours) or allow better integration with intermittent sources such as solar or wind.

Table 1. Fuel Gas Specifications for Gas Turbines

Contaminant	Gas Turbine
Total non-particulates	Not available
Total sulfur (H_2S, COS, etc.)	750 ppmv fuel gas 20 ppmv for Selective Catalytic Reduction (SCR)
Total halides (Cl, F, Br)	5 ppmv fuel gas
Total fuel-nitrogen (NH_3, HCN)	Fuel-bound nitrogen 200–400 ppmv
Total alkali metals (Na, K, Li vapor and solid phases)	100 ppbv fuel gas
Volatile Metals (V, Ni, Fe, Pb, Ca, Ba, Mn, P)	20 ppbw Pb[a] 10 ppbw Va 40 ppbw Ca[a] 40 ppbw Mg[a]
Water	Not available
Total hydrocarbons (C_1 basis)	Not available
Oxygen	Not available
Carbon dioxide	Determined by required carbon capture
Carbon monoxide	Determined by required carbon capture
Formaldehyde	Not available
Formic acid	Not available
Particulates	0.1–0.5 ppmw fuel gas

[a] Specification for Fuel Gases for Combustion in Heavy-Duty Gas Turbines, GEI 41040G, GE Power Systems, Gas Turbines, January 2002. http://www.netl.doe.gov/technologies/ coalpower/turbines/ refshelf/GE%20Tur bine%20Fuel%20Specs.pdf

3.2. Comparison of Current and Future Technology

At the present time, only a small number of IGCC plants operate on a commercial scale worldwide, and none use membrane separation technology. IGCC plants using current technology without CCS have not flourished due to higher capital costs than pulverized coal (PC) plants, leading to a higher cost of electricity. However, IGCC plants have lower costs to capture CO_2, leading to lower costs of electricity than PC when CCS is incorporated.[3] NETL has also conducted further analysis which shows the technical, environmental, and economic benefits of incorporating advanced technologies (including membrane separation, warm gas cleanup (WGCU), advanced turbines, advanced coal pumps, and advanced oxygen separation) into IGCC plants with CCS. Select results from this analysis can be seen in Table 2. All of these IGCC cases involve capture and sequestration of 90 percent of total $_{CO2}$.

NETL's analysis shows that integration of advanced hydrogen separation membranes offers the single biggest cost savings and incremental improvement in efficiency of any single advanced IGCC technology under development. This efficiency improvement is nearly equal to the incremental improvement from moving two generations forward in turbine technology, from the current 7FA turbine to one generation past the advanced F-series turbines.

Table 2. Summary of IGCC Cases with CCS

	Units	CASE 1 – Reference	CASE 2 – WGCU/Selexol	CASE 3 – WGCU/ Hydrogen Membrane	CASE 4 – Hydrogen Membrane/ Advanced Turbine
Technology readiness goal	–	Current	N/A	2015[a]	N/A
Efficiency	%	30.4	33.3	36.2	40.0
Total Plant Cost	$/kW	2,718	2,425	2,047	1,683
Levelized Cost of Electricity	cents/kWh	11.48	10.00	8.80	7.36

Source: DOE/NETL, *Current and Future Technologies for Gasification-Based Power Generation: Volume 2*, November 2009 [a] See Section 4 for additional hydrogen membrane goals, milestones, and technical targets.

NETL's IGCC reference case, shown schematically in Figure 3, produces electric power and sequesters CO_2 using gasification and separation technologies considered state-of-the-art in 2003. The plant has an 80 percent capacity factor, and uses cryogenic air separation to produce oxygen, which is fed to single-stage slurry feed gasifiers with radiant-only syngas coolers. The plant uses two trains of water quench and a sour water gas shift to produce extra hydrogen, followed by a two-stage Selexol process to remove CO_2 and acid gases. Sulfur is then recovered using Claus technology, and the hydrogen is burned in a 7FA hydrogen turbine. CO_2 is compressed and transported offsite for subsequent sequestration. This plant configuration leads to a total plant cost of $2,718/kW and has an efficiency of 30.4 percent, leading to a 20-year levelized cost of electricity of 11.48 cents/kWh.

To show the incremental benefits of each advanced IGCC technology, NETL developed a series of cases based on this reference plant which added new technologies one at a time. The first incremental case (Case 2) has a higher capacity factor than the reference case at 85 percent, and as seen in Figure 4, utilizes similar air separation and gasification equipment to produce syngas. The cold gas cleanup and two-stage Selexol acid gas removal system is replaced with a warm gas cleanup system and uses only a single-stage Selexol system to remove CO_2. This advanced cleanup system includes removal of sulfur, hydrochloric acid, ammonia and mercury. The clean fuel gas is then sent to an advanced F-class turbine for power generation. This plant configuration leads to a total plant cost of $2,425/kW and has an efficiency of 33.3 percent, leading to a 20-year levelized cost of electricity of 10.00 cents/kWh.

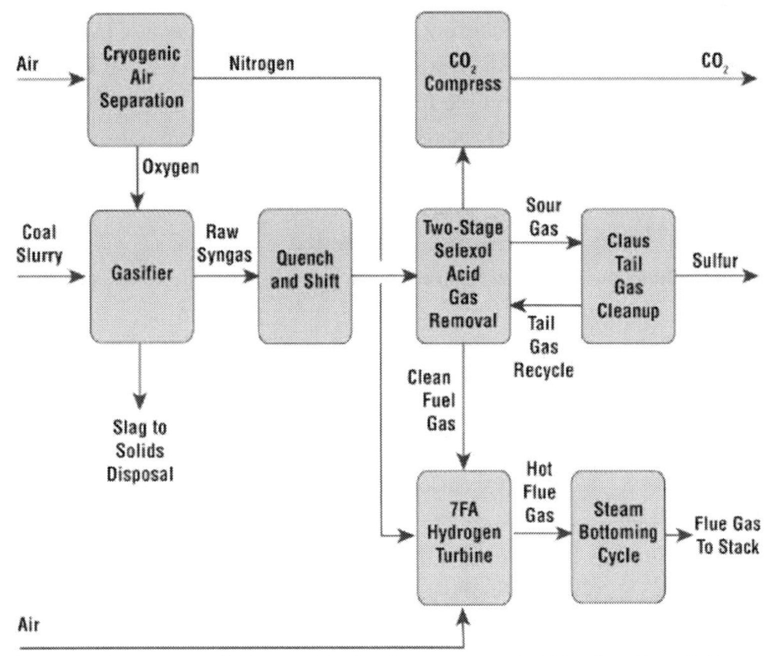

Source: DOE/NETL, Current and Future Technologies for Gasification-Based Power Generation: Volume 2, November 2009.

Figure 3. Case 1: Reference Case for Carbon Capture Plant, 80% Capacity Factor.

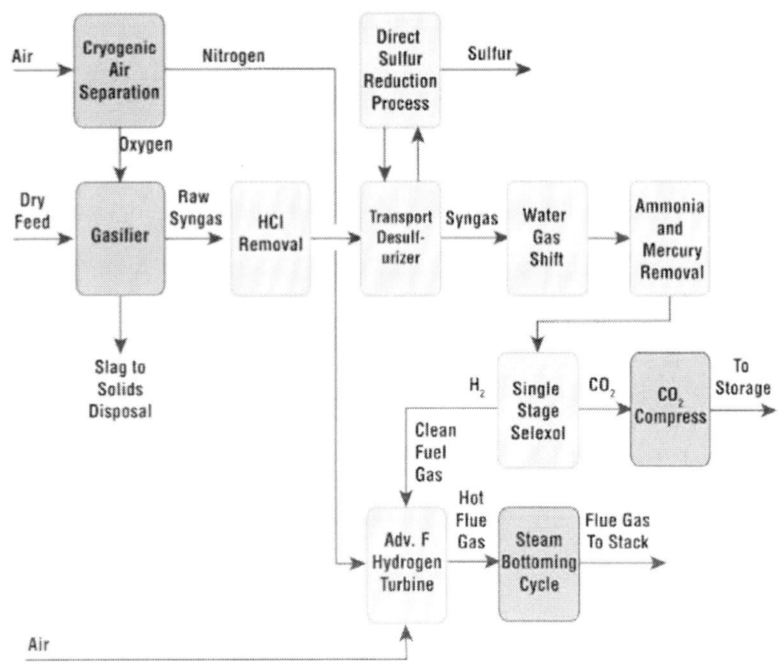

Source: DOE/NETL, Current and Future Technologies for Gasification-Based Power Generation: Volume 2, November 2009.

Figure 4. Case 2: Reference Case Plus Advanced F Turbine, 85% Capacity Factor, WGCU and Selexol Separation.

Case 3, shown in Figure 5, incorporates advanced hydrogen separation membranes, currently under development by the Hydrogen from Coal RD&D Program. In this case, the single-stage Selexol CO_2 separation is replaced by a hydrogen membrane, while other upstream and downstream processes, including the use of warm gas cleanup, remain the same. The membrane concept used for this analysis is assumed to meet the Program's 2015 Technical Target for hydrogen flux, hydrogen purity, operational flexibility, and cost (see Section 4 for details). When using hydrogen separation membranes, hydrogen gas permeates through the membrane and is then swept with nitrogen gas from the air separation unit to keep a low partial pressure of hydrogen, which enhances the hydrogen flux in the membrane. The CO_2 remains on the non-permeate side, which allows it to remain at a high pressure and reduces the amount of energy required to compress it to the pressures required for sequestration applications. Additionally, efficiency improvements are seen due to reduction of extra equipment associated with the Selexol process. Overall, replacing the one-stage Selexol separation system of Case 2 with a hydrogen separation membrane system improves plant efficiency by 2.9 percent.

This plant configuration leads to a total plant cost of $2,047/kW and has an efficiency of 36.2 percent, leading to a 20-year levelized cost of electricity of 8.80 cents/kWh. This overall capital cost for the plant using WGCU and membrane separation is $378/kW lower than the WGCU and Selexol separation plant, with $189/kW being from the reduced cost of the hydrogen separation unit, and $49/kW due to decreased compression requirements for CO_2. There is also a reduction in turbine operating costs due to elimination of the syngas expander. The remaining savings are primarily due to the increased efficiency of the plant.

When combining the effects of improved efficiency and lower capital costs, using a hydrogen membrane system lowers the overall cost of electricity by 12 percent compared to the one-stage Selexol system in Case 2.

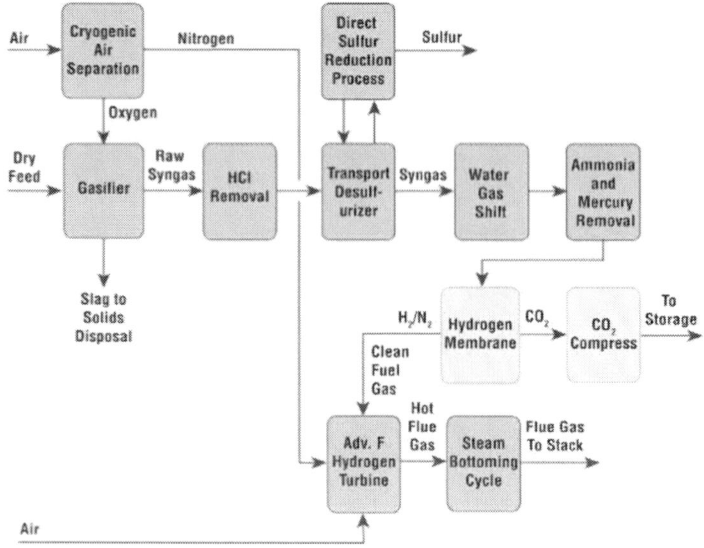

Source: DOE/NETL, Current and Future Technologies for Gasification-Based Power Generation: Volume 2, November 2009.

Figure 5. Case 3: Reference Case Plus Advanced F Turbine, 85% Capacity Factor, WGCU, and Hydrogen Separation Membrane.

Finally, the NETL analysis showed the cumulative benefits of employing all advanced IGCC technologies currently being researched. This case (Case 4), shown in Figure 6, involved replacing the cryogenic air separation technology with an ion transport membrane to produce oxygen for the gasification reactor. Additionally, the Advanced "F" turbine was replaced with the most advanced turbine technology currently being developed by DOE, referred to as an AHT-2 turbine. When combined with a 90 percent capacity factor, this plant configuration leads to a total plant cost of $1,683/kW and has an efficiency of 40.0 percent, leading to a 20-year levelized cost of electricity of 7.36 cents/kWh.

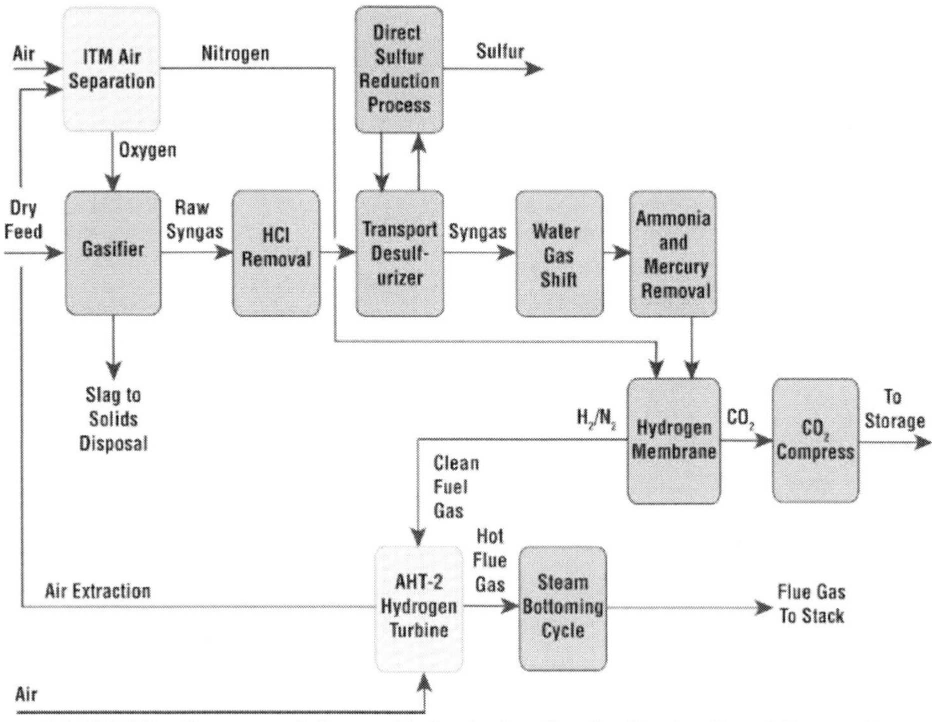

Source: DOE/NETL, Current and Future Technologies for Gasification-Based Power Generation: Volume 2, November 2009.

Figure 6. Case 4: Reference Case Plus Advanced AHT-2 Turbine, 90% Capacity Factor, WGCU and Hydrogen Separation Membrane, Membrane Air Separation.

4. TECHNICAL PLAN

The Hydrogen from Coal Program RD&D Plan supports the overall Office of Clean Coal's goals of improved energy security, reduced GHG emissions, high tech job creation, and reduced energy costs through joint public and private RD&D of advanced and novel hydrogen-related technologies for the future low-carbon energy system.

As successes are achieved, this RD&D program will improve existing technology and make available new, innovative technology that can produce affordable hydrogen from coal with significantly reduced or near-zero emissions. These technologies will be discussed in detail in this section, and are further broken down into specific technological areas. Each technology will include goals and milestones as well as technical targets, where appropriate. These goals and milestones are continually validated and/or updated based on the changing market and technical needs and the progress being achieved with individual projects.

FE's Office of Clean Coal and its implementing arm, the National Energy Technology Laboratory's Strategic Center for Coal, have R&D activities on coal gasification, advanced turbine development, fuel cells, and carbon sequestration technologies to improve the efficiency of power production and to reduce the environmental impact of coal use. These efforts are not part of the Hydrogen from Coal Program, but instead are technologies under development in other OCC and NETL programs. Therefore, R&D efforts in these research areas represent associated rather than direct elements of the Hydrogen from Coal Program.

The focus of the Hydrogen from Coal Program RD&D efforts is on those technologies that employ hydrogen membrane separation, sorbent-based concepts, and advanced concepts such as co-production and process intensification. Today's unit operations for producing hydrogen as part of an IGCC power plant are effective but also are expensive and energy-intensive.

Novel technologies could be developed that combine multiple processes into one step (i.e., process intensification technology), be better integrated with highly efficient warm gas cleanup systems, and/or remove impurities such as sulfur and CO_2 into one stream that can be jointly sequestered. The resulting benefits would include higher process efficiency and lower costs.

4.1. Goals and Milestones

The goals of the Hydrogen from Coal Program represent aggressive cost and efficiency savings realized through meeting the Program's technical targets laid out in Section 4.4, and are as follows:

- Prove the feasibility of a 40 percent efficient, near-zero emissions power facility that uses membrane separation technology as well as other advanced technologies to reduce the cost of electricity by at least 35 percent (relative to a base case IGCC with CCS using currently available technologies).
- Develop hydrogen production and processing technologies that will contribute approximately 2.9 percent in improved efficiency and 12 percent reduction in cost of electricity to the 40 percent efficient near-zero emissions power facility.

Technical Targets for the program include the following:

- Develop hydrogen membranes by 2015 that can provide hydrogen fuel that meets gas turbine specifications from syngas under warm gas cleanup conditions at a flux level of 300 SCFH/ft^2 at pressures that will reduce compression costs for H_2 and CO_2, and at a membrane of less than \$100 per ft^2.
- Develop advanced concepts for syngas processing in gasification-based electric power generation systems that will significantly reduce complexity and cost, improve plant and fuel utilization, and/or improve efficiency, to produce H_2 for gas turbine fuel and CO_2 for sequestration. The program builds on expected RD&D successes in

associated programs within FE. After initial success by multiple advanced membrane systems on a laboratory scale, the program has begun to give additional focus to the scale-up of these systems to pilot scale. Figure 7 shows the proposed developmental schedule.

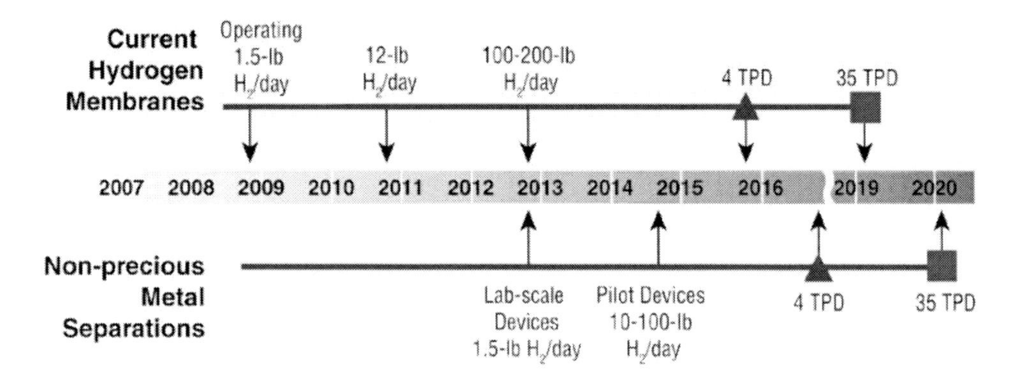

Figure 7. Hydrogen from Coal RD&D Program Milestones.

Current Hydrogen Membranes

End of 2008: Operating at 1.5 lb H_2/day

End of 2010: Operating at 12 lb H_2/day

End of 2012: Operating at 100–200 lb H_2/day

End of 2015: Scale-up membrane module to 4 tons H_2/day production for component testing Beginning of 2019: Scale-up membrane module to 35 tons H_2/day production for integrated testing

Non-precious Metal Separations

End of 2012: Operating lab-scale devices at 1.5 lb H_2/day

End of 2014: Operating pilot-scale devices at 10–100 lb H_2/day

End of 2016: Scale-up non-precious metal membrane module to 4 tons H_2/day for component testing Beginning of 2020: Scale-up non-precious metal membrane module to 35 tons H_2/day for integrated testing

4.2. Activities

Table 3 lists the RD&D activities under investigation by the Hydrogen from Coal Program.

Both FE and NETL have acquired extensive research experience in all aspects of producing hydrogen from coal through their participation in the Advanced Power Systems, Ultra-clean Fuels, and Advanced Research programs. Exploratory research previously sponsored by FE and NETL has pioneered studies on palladium-copper alloy membranes; tested novel membranes with regard to flux, durability, and impurity resistance; evaluated WGS kinetics and advanced reactor systems; and explored new concepts and fundamental studies on novel separation systems.

Table 3. Relevant Current R&D Program Activities

Category	Technology
Advanced Membrane Separation Systems	• Palladium and palladium alloy membrane reactors • Cermet membranes
	• Non-precious metal membranes
	• Microporous membranes (e.g., carbon molecular sieves)
Sorbent-based Concepts	• Chemical looping (e.g., iron-calcium cycle process to produce hydrogen and sequestration-ready CO_2) • Hydrogen sorbents
Advanced Concepts	• Process intensification (e.g., WGS membrane reactors) • Co-production concepts

4.3. Technologies

The technologies within the Hydrogen from Coal Program are provided in the list below and discussed in further detail in the denoted section of the Plan.

4.3.1 Advanced Membrane Separation Systems

4.3.2 Sorbent-based Concepts

4.3.3 Advanced Concepts

4.3.1. Advanced Membrane Separation Systems

Modern gasification and WGS technologies produce synthesis gas (syngas), a mixture of H_2, CO, CO_2, and other chemical compounds. There are several gas separation technologies that could separate constituents of the syngas. The Hydrogen from Coal Program seeks to develop technologies to improve the separation of H_2 and/or conversely, CO_2 from synthesis gas streams, that will reduce capital and operating costs and improve thermal efficiency and environmental performance. Membranes to separate O_2 from air are being developed in the OCC Gasification Technologies Program.

Current hydrogen recovery methods typically employ pressure swing adsorption (PSA), cryogenics (e.g., Selexol), or polymer membrane systems. Each of these technologies has limitations. PSA typically recovers less of the feed-stream hydrogen and is limited to modest temperatures. Cryogenics is generally used only in large-scale facilities with liquid hydrocarbon recovery because of its high capital cost.

Current polymer membrane systems are susceptible to chemical damage from H_2S and aromatics, as well as having limited temperature tolerance.

There are significant opportunities to make advancements in these separations with the development of various types of advanced membranes that can separate hydrogen from CO_2. Much of the work will develop technology modules that are efficiently integrated into the plant systems and optimized with the temperature and pressure requirements of the plant and the specifications of the hydrogen for turbine use.

Advancements in hydrogen membrane separation technologies have the potential to reduce costs, improve efficiency, and simplify hydrogen production systems. Desirable characteristics of separation membranes include high hydrogen flux at low pressure drops; production of H_2 and CO_2 at high pressures; tolerance to contaminants, especially sulfur and CO; low cost; and operation at system temperatures of 250–500°C. Many current hydrogen membrane technologies are at the research phase, but because of their characteristics, they have the potential to provide hydrogen purity above 99.99 percent. Scale-up of these membrane technologies remains a key focus of the Hydrogen from Coal Program, and this activity is ongoing.

Membranes can be classified as organic, inorganic, or hybrid (a mixture of organic and inorganic materials). Within each of these classes, membranes can be characterized based on their properties. The Hydrogen from Coal Program currently is focused on microporous and metallic membranes, which include pure metal and hydrogen-permeable ceramic-metal membranes (i.e., cermets). The program previously included dense ceramic membranes (and also non-hydrogen permeable cermets) as part of its research activities. Dense ceramic membranes separate hydrogen from mixed gas streams by transferring hydrogen ions (i.e., protons) and electrons through the membrane matrix. These membranes have interesting characteristics such as high-temperature operation, mechanical stability, and very high hydrogen selectivity. However, hydrogen flux rates are low at gasifier effluent and gas clean-up technology operating conditions, which would significantly increase the cost of the separation module. Therefore, the program has de-emphasized RD&D activities on dense ceramic membranes.

With the refocus in the activity for increased power production, emphasis will be placed on higher flux and lower purity hydrogen by use of lower cost metal membranes. Membrane separation technology will continue to be the paramount research activity for hydrogen separation within the Program. However, a major refocus will occur based on systems analysis and activity requirements, which will include review of current membrane activities to define their relevance to the modified objectives.

A brief characterization of the current membrane technologies being developed by the Hydrogen from Coal Program is provided below. Other membranes are not precluded, provided that they show potential to meet the technical targets and assist the Hydrogen from Coal Program to meet its goals and milestones.

- **Metallic Membranes** – These membranes include pure metal or metal alloys, and hydrogen permeable cermets. The flux for these membranes is proportional to the differences of the square roots of the partial pressures across the membrane. Because of the transfer mechanism involved, 100 percent pure hydrogen can be recovered. A description of the three metallic membrane sub-types is provided below.
  - **Pure metal and metal alloy membranes.** Pure metal and metal alloy membranes transport gaseous hydrogen via an atomic mechanism whereby the metal or metal alloy, usually made with palladium (Pd), dissociates the molecular hydrogen into atoms that pass through the Pd metal film, and the atoms recombine into hydrogen molecules on the other side of the membrane. These metallic membranes typically comprise metal composites, thin Pd, or a

Pd-alloy metal layer supported on an inexpensive, mechanically strong support. The hydrogen diffuses to the metal surface where dissociative chemisorption occurs, followed by absorption into the bulk metal and diffusion through the metal lattice and recombination into molecular hydrogen at the opposite surface, and finally diffusion away from the metal membrane. These micro-thin metallic films are poisoned by gaseous impurities like sulfur compounds and carbon monoxide, and at high temperatures they may undergo phase changes that significantly reduce the hydrogen flux. Alloying with other metals like copper and silver reduces this phase change propensity.

- **Hydrogen permeable cermets.** In the second type of metallic membrane, a dense mixed conducting ceramic matrix phase is combined with a hydrogen-permeable metallic second phase. This metallic phase, which is composed of a hydrogen permeable metal or metal alloy, functions in the same way as the metallic membranes described previously. In this mixed membrane, the mechanism of hydrogen transfer is a combination of proton and electron conductivity in addition to atomic hydrogen transfer. However, atomic hydrogen transfer is orders of magnitude greater than the contribution of proton and electron conductivity, and thus the overriding mechanism in estimating the flux. Therefore, the flux for this membrane is more closely related to that of metallic membranes (i.e., represented by the difference in the square roots of the partial pressures). The membranes can operate at temperatures in the range of 400–600 °C, and can produce 100 percent pure hydrogen because of the transfer mechanism involved. These ceramic/metal composites offer the potential to overcome many of the limitations of metal membranes. This includes inhibition of phase change and increased tolerance to impurities in the synthesis gas.
- **Advanced non-precious metal membranes.** Within the context of the current program – non precious metal membranes will play an important role since the purpose is to find membranes which may not need as stringent hydrogen purity requirement, but would concentrate on higher flux and lower cost domestic materials. Materials will have to be evaluated to define the preferred hydrogen purity and fluxes based on the required conditions of temperature and pressure. It is expected that the types of non precious metal membranes would be similar to the membrane types identified above.

- **Microporous Membranes** – These membranes separate molecules through Knudsen diffusion, molecular sieving, surface flow, or a combination of these transport mechanisms. Flux increases linearly with increasing pressure, and there is usually a flux increase with higher temperatures. Materials such as ceramics, graphite, or metal oxides can be used in making these membranes. These materials can be stable in harsh operating environments. The pores in the membrane may vary between 0.5 nanometers (nm) and 5 nm. These membranes are characterized by higher fluxes and lower hydrogen purities which may have increased relevance with the re-focused emphasis on purities needed for use in hydrogen turbines.

Figure 8 shows the current performance characteristics for Pd or Pd alloy-based metallic or composite membranes under development by FE and NETL. Some of these membranes are approaching the desired flux rates of about 300 SCFH/ft^2 at 100 psi ΔP hydrogen partial pressure and the desired operating temperature range of 250–550°C. However, these flux measurements have mostly been made in pure gases (H_2 and He), and the addition of H_2S has been shown to reduce hydrogen flux rates in many test membranes. Other characteristics, such as desired durability, have not yet been demonstrated.

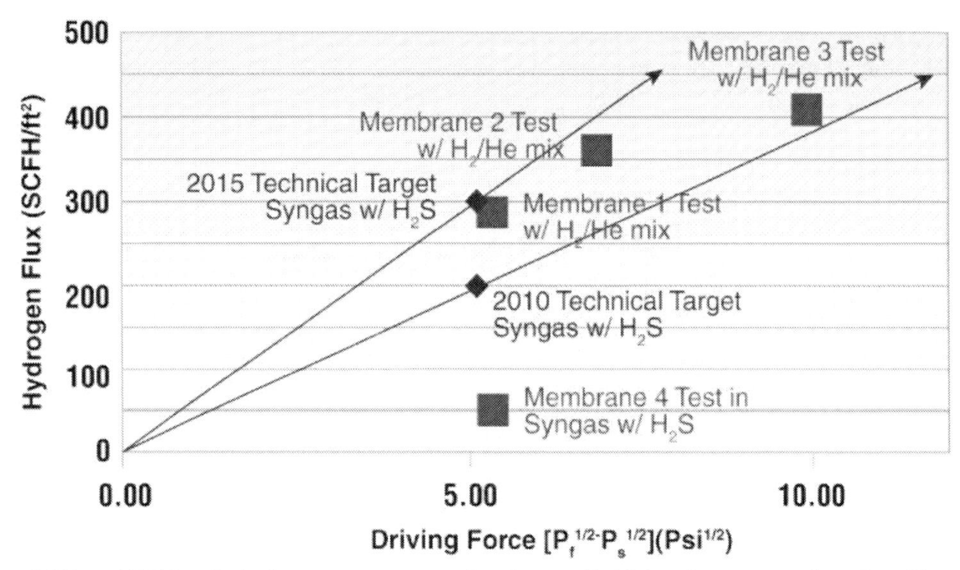

Note: 2010 and 2015 technical targets are extrapolated over all driving force ranges based on Sievert's Law.

Figure 8. Current Status of the Membrane Development Activities Sponsored by FE and NETL.

Membrane flux is dependent upon the partial pressure of hydrogen, and the relationship between the two differs depending upon the type of membrane. Specifically, microporous membranes exhibit a flux that is directly proportional to the hydrogen partial pressure differential across the membrane. In Pd or Pd alloy-based metallic or composite membranes, the flux is proportional to the difference in the square roots of the partial pressures or the natural log of the partial pressure gradient according to Sievert's Law. In dense ceramic and non-hydrogen permeable cermets, flux is proportional to the natural log of the pressure gradient across the membrane, based on the Nernst potential.

In addition to hydrogen partial pressure, other operating conditions such as temperature and quality of the feed stream can also influence hydrogen flux. Membrane attributes such as durability, cost, tolerance to contaminants, hydrogen recovery, and purity are also important factors in development of robust membranes that can be integrated into coal-based hydrogen production facilities.

Because of the complexities of membrane performance, it is important to set desired technical targets near the expected operating conditions. In the case of hydrogen from coal technologies, hydrogen separation membranes are expected to operate with at least 50 psi hydrogen partial pressure on the permeate side and a hydrogen partial pressure ΔP of 100–300 psi.

To ensure that the many types of membrane technologies and concepts being researched are on a consistent basis, NETL has developed a standardized testing protocol for hydrogen separation membranes. The testing protocol[4] serves to accomplish a number of objectives with respect to the Hydrogen from Coal Program technical targets. These objectives involve clearly stating expectations to contractors, determining the effectiveness of each membrane on a common basis, and assessing the membrane's compatibility with current gasification operation conditions.

4.3.2. Sorbent-based Systems

Chemical looping (CL) is a process that takes advantage of the ability of some materials to adsorb and desorb oxygen in a cyclic manner, similar to the way that hemoglobin in the blood stream carries oxygen from the lungs to muscle cells. CL is a transformative technology capable of converting carbon-based feedstocks into energy more efficiently than existing combustion or gasification routes, and inherently produces a sequestration-ready CO_2 stream. For gasification applications, the basic concept is to supply oxygen to the gasification process from a metal that is oxidized in air, and then chemically reduced with the fuel. The metal oxidation/reduction process is repeated in a cyclic "loop," hence the name chemical looping.

Sorbent-based separation systems operate by taking advantage of the physical or chemical properties of certain materials at varying temperatures or pressures. As shown in Figure 9, syngas is fed from the gasifier on to a bed of hydrogen adsorbing material at high temperatures and pressures. Hydrogen preferentially adsorbs onto the sorbent surface, while CO_2 and other syngas components remain in the bulk gas and exit as a permeate. The sorbent is then exposed to lower pressure (i.e., PSA) or temperature (i.e., thermal swing adsorption, or TSA), at which point the sorbent releases the hydrogen on its surface. Two separate sorbent columns are used in this process in order to maintain a continuous system. One column adsorbs hydrogen from syngas while the other is regenerated to remove the majority of trapped hydrogen and a purge gas is used to release the remainder.

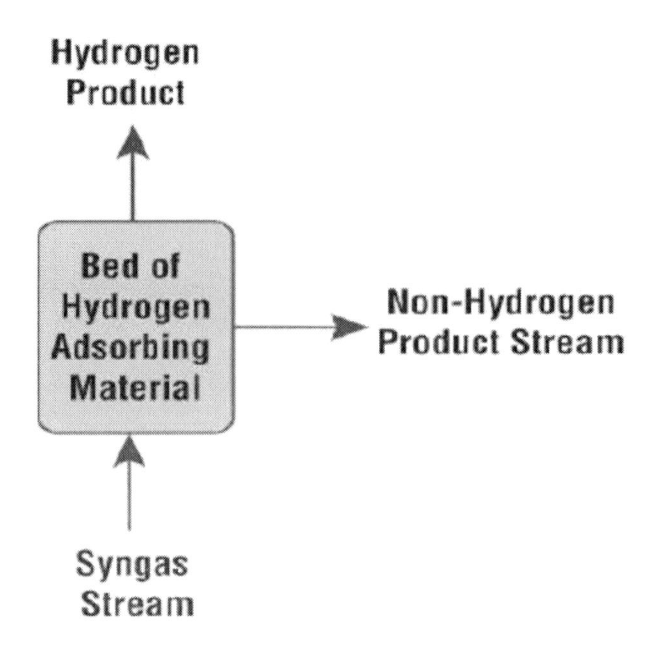

Figure 9. Hydrogen Sorbent Separation Process.

For a TSA-based system, extra bed heating intervals would need to be added to the process, and means would need to be included for introducing heat without diluting the product hydrogen. Heat exchange might be provided by tubes in the bed or transfer of hot solids as in chemical looping processes.

Additionally, the FE Carbon Capture Program is investigating sorbents which preferentially adsorb CO_2. These sorbents may be of particular use in a hydrogen production process, as removing CO_2 from syngas would leave a highly concentrated hydrogen product behind.

4.3.3. Advanced Concepts

The Hydrogen from Coal Program is investigating advanced concepts through process intensification. Process intensification is the concept of developing novel technologies that, compared to current technology, bring about dramatic improvements that lead to more compact, energy efficient, and lower cost technologies. As related to hydrogen production from coal, these concepts could be a "one-box" process that combines synthesis gas cleanup, the WGS reaction, and hydrogen separation. Others include new process control methods or novel concepts that integrate alternative energy sources into the hydrogen from coal production facility. These advanced concepts will require long-term research efforts before they are ready for larger-scale development, but could significantly improve the production of hydrogen from coal.

The WGS membrane reactor (WGSMR) concept is of particular interest as an example of process intensification. The WGSMR concept has been around for some time, dating back to at least the early 1990s. Figure 10 below shows the concept. The membrane and WGS catalyst is combined in such a way as to transport hydrogen through the membrane — away from the feed side of the membrane where the WGS catalyst is located — taking advantage of Le Chatelier's Principle to allow maximum conversion of CO to H_2 in a single reactor.

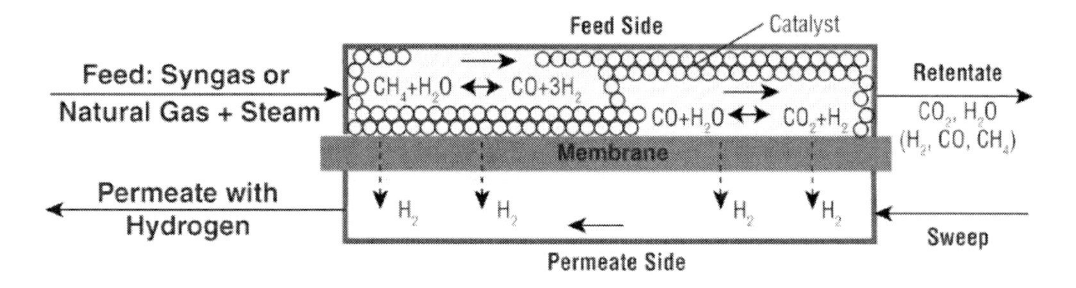

Figure 10. Advanced WGS Membrane Reactor Concept.

The conditions for WGS in a membrane reactor would be very different from those encountered in a conventional shift reactor. In particular, at higher conversion rates, the partial pressure of steam would be low and the gas phase would be predominantly either CO_2 or H_2, depending on which species was not removed through the membrane. The catalysts used in conventional WGS reactors have not been extensively studied at these conditions, and it is not known whether they will be suitable for use in membrane reactors.

In either case, membrane reactors are being considered for this application with the expectation that, among other advantages, using a membrane reactor would not require cooling the gasifier product as much as is required in a conventional shift reactor. WGS is a reversible, exothermic reaction, and consequently the conversion is limited at high temperatures by thermodynamic equilibrium. The use of a highly permeance-selective membrane reactor would avoid this problem, driving the reaction to high conversion rates at elevated temperatures by selectively removing either H_2 or CO_2 (Le Chatelier's Principle). While this conceptually removes the limitation on conversion at high temperature, it is unknown what effect it will have upon the rate of reaction or mass transfer limitations. Conducting the WGS reaction over commercial, high-temperature iron oxide catalysts is known to be inhibited by the CO_2 reaction product. While the kinetics are not inhibited by the H_2 reaction product, it is not known whether the active (Fe_3O_4) state of the catalyst can be maintained in the situation where CO_2 is removed and where an excess of steam is neither needed nor desired. In a membrane reactor, one or the other of these compositional regimes will be encountered.

Another advanced concept being investigated by the Hydrogen from Coal Program is that of coproduction, in which part of the fuel gas feed intended for turbine combustion is reacted to form other high-value products (e.g., liquid fuels, chemicals, hydrogen, or substitute natural gas) for export. Varying the use of syngas to produce high-value products offers the potential to maximize gasifier utilization during periods of low daily or seasonal demand for electric power. Additionally, depending on the value of the secondary product, the economics of the plant can be improved. This may result in a lower cost to produce electricity.

4.4. Technical Targets

The technical targets in this RD&D Plan, unless otherwise indicated, represent the status of the specific technologies after completion of R&D, but prior to large-scale demonstration of the technologies. These technologies will be validated in modules at facilities that can accommodate similarly scaled engineering modules, as detailed in the Gantt chart in Figure 7. As a point of reference, the status of the technologies is provided in the technical target tables.

It is important to understand the composition of the synthesis gas exiting the gasifier when developing the targets for contaminant levels for both WGS and hydrogen separation technologies. Table 4 shows the contaminant levels in raw and cleaned synthesis gas from Illinois #6 bituminous coal. Additionally, the FE Gasification Technologies Program goals for synthesis gas cleanup are also shown. It should be noted that raw synthesis gas composition will vary by coal type; therefore, lower sulfur coals could have raw synthesis gas sulfur levels that are much lower than those shown in Table 4, perhaps as low as 700– 1,200 ppmv based on some studies.[5]

As Table 4 shows, most major contaminants can be reduced to very low levels through conventional synthesis gas cleaning technologies, and achieve the goals of the Gasification Technologies Program. If that program's cleanup goals were achieved then conventional WGS, rather than a sour gas shift to produce additional hydrogen, would be preferred. An alternative for advanced systems would be to assume that sulfur can be controlled to about 100 ppmv by use of warm gas cleanup, but without substantial removal of other contaminants such as ammonia, HCl, or mercury below that seen in the gasifier effluent. These contaminants, along with CO_2, would be simultaneously removed by the separation device and thereby significantly reduce the cost to produce hydrogen fuel. The 2015 targets for WGS and membrane separation assume tolerances for the identified contaminants consistent with this methodology. However, under the current cleaned synthesis gas composition, sulfur levels in the form of H_2S are considerably higher than the Gasification Technologies Program goal and would require a sour gas shift that might affect advanced hydrogen membrane separators.

Table 4. Contaminant Levels in Raw and Cleaned Synthesis Gas using Conventional Cleaning Technologies and FE Gasification Program Goals for Synthesis Gas Cleanup

Contaminant	Units	Raw Synthesis Gas Composition[a]	Cleaned Synthesis Gas Composition[a]	FE Gasification Program Goals[b]
H_2S	ppmv	9,524	102	0.04
NH_3	ppmv	675	0.4	1,000
HCl	ppmv	425	~0	< 1
Hg	ppbv	3	0.3	< 1

[a] Novel Gas Cleaning/Conditioning for Integrated Gasification Combined Cycle: Volume I – Conceptual Commercial Evaluation, Siemens Power Generation, Inc. and Gas Technology Institute, under DOE Contract DE-AC26-99FT40674, December 2005.

[b] Tennant, J., "Gasification: Ultra Clean & Competitive," DOE/NSF EPSCoR Conference 2005, Morgantown, WV, June 2005.

To reiterate, it is also expected that efficiency requirements for advanced concepts (i.e., process intensification) will require "warm gas cleanup," which will have higher levels of sulfur than conventional cleaning. In addition, these advanced concepts may require that the effluent from the gasifier be processed without major cleaning. Therefore, the advanced concepts may require WGS and membrane separation with sulfur, ammonia, and chloride tolerances that are found in the raw gasifier effluent. This suggests that the WGS and membrane tolerances to contaminants in the synthesis gas require a better quantitative definition and may be different depending on the specific research approach being pursued.

4.4.1. Hydrogen Separation Technical Targets

The performance criteria for successful incorporation of membrane separation reactors into hydrogen from coal configurations are shown in Table 5. Although high flux rates and low cost are the key parameters, there also are other critical criteria that must be satisfied. Ideally, the temperature of operation should be in a range compatible with warm synthesis gas cleaning technologies.

The technical targets for hydrogen membranes relate to hydrogen from coal technology in which ΔP will be around 100 psi and the membrane will require resistance to contaminants (CO and H_2S).

The targets are not independent, but are part of an integrated set of requirements that are necessary for hydrogen separation membranes to be operated in a coal-to-hydrogen production plant using advanced technologies and capable of producing hydrogen with high efficiency and low cost. A typical plant for hydrogen production from coal using advanced separation membranes would use an advanced gasification system, such as an entrained flow, transport or circulating fluidized bed gasifier; a warm or hot gas cleanup system to remove ash, sulfur compounds, and other contaminants such as mercury; WGS catalysts to maximize hydrogen production and reduce carbon monoxide content in the gas; and the advanced hydrogen membrane to separate product hydrogen and capture carbon dioxide for sequestration. The plant might also incorporate membrane separation to separate oxygen from air for gasification. All of these advanced systems are important to achieving the greatest efficiency and lowest cost of production. Hydrogen membrane separation is an enabling technology that allows maximum flexibility to integrate with other advanced gasification and gas cleanup technologies to achieve high efficiencies, low GHG emissions, and low costs in gasification-based energy systems.

Table 5. Hydrogen Separation Technical Targets

Performance Criteria	Units	Current Status [b] (Pd-based metallic or composite)	2010 Target	2015 Target
Flux [a]	$ft^3/hour/ft^2$	~200–300	200	300
Temperature	°C	300–500	300–600	250–500
S tolerance	ppmv	tbd	20	>100
Cost	$\$/ft^2$	tbd	100	<100
WGS activity	–	N/A	Yes	Yes
ΔP Operating capability[c]	psi	tbd	Up to 400	Up to 800 to 1,000
Carbon monoxide tolerance	–	Some	Yes	Yes
Hydrogen purity	%	>99.5	99.5	99.99
Stability/durability	Years	0.9 (tested)	3	5

[a] For 100 psi ΔP (hydrogen partial pressure basis). ΔP = total pressure differential across the membrane reactor.

[b] Detailed status of hydrogen membrane separation can be found in Figure 8.

The flux and cost targets, as well as the lifetime target, represent conservative goals that will allow designers to keep overall plant costs down while allowing flexibility in configuration of membrane separation systems in conjunction with other advanced technologies (WGCU) to achieve maximum efficiencies and GHG reductions. Very high product purity is required for use of hydrogen in fuel cells. (Note: while the purity of hydrogen product from advanced separation membranes is very high, some hydrogen stays in the gas stream with the CO_2 and multiple membrane stages are needed to recover this hydrogen). The flux targets were established by expert estimation of the degree of performance improvement that is judged to be achievable by membranes research in balance with what is deemed to be acceptable total membrane surface areas and costs in a plant. As an example, the 2015 flux target of 300 $SCFH/ft^2$ of pure hydrogen (density of .005 lb/SCF) is approximately 16 kg/day per membrane square foot. For a production plant producing 158 million SCF per day of hydrogen (equivalent to approximately 625 MW or roughly 8,500 barrels per day of liquid fuels), this implies a membrane area for a single stage membrane system of 22,500 ft^2. More detailed designs estimate that this area will need to increase by a factor of 2 or 3 to allow for multiple separation stages for high hydrogen recovery. The resulting membrane area of ~ 50,000 ft^2 can be achieved by arrays of separation tubes or other designs. If the membranes are tubular, on the order of 5,000 tubes might be needed.

Clearly total hydrogen membrane system costs are inversely proportional to achievable flux levels.

The 2010 Target for membrane temperature (Table 5) allows the hydrogen membrane to be installed and operated in or near a low temperature or high temperature (clean gas) WGS reactor, but the gas must be cleaned of most sulfur compounds first. A higher upper temperature is allowed for the 2010 target because these membranes typically have larger flux at higher temperatures. The 2015 target actually reduces the temperature range slightly because WGCU and WGS systems operate in a lower range. However, the 2015 Target allows for higher sulfur levels (see below) and thus more flexibility in configuration, especially with warm gas cleanup systems. Warm gas cleanup systems offer significant improvements in efficiency because less heat exchange is necessary in the gas processing systems.[6] Warm gas cleanup systems that match the temperature and pressure conditions for the WGS reaction and the advanced hydrogen membranes will minimize the amount of heat lost to inefficiencies in heat exchanges and will improve overall plant efficiency. Warm gas cleanup and WGS reactions typically operate in the 250–500°C range, which was selected as the optimal separation membrane temperature range. Shift reactors must be used to convert water and CO in the gas into additional H_2 and CO_2 in order to maximize the H_2 content for separation and make CO_2 available for sequestration. Raw gas shift reactors using sulfur resistant catalysts can be applied to the gas before desulfurization at temperatures of 250 to 550°C. Alternatively, clean synthesis gas after sulfur removal can be reheated to 315–370°C for WGS reactors that produce additional hydrogen through the catalytic reaction of CO with H_2O to form CO_2 and H_2.

Sulfur tolerance is one of the most difficult elements of the technical targets to achieve. Achievement of membrane tolerance to the 2010 target of 20 ppm level of sulfur species would allow the membrane to be used in a gas stream containing low levels of sulfur, typical of cold gas desulfurization or very effective warm gas cleanup. However, as hydrogen permeates through the membrane, the concentration of sulfur on the non-permeate side is increased due to the decreased total volume. This causes some membrane components to be exposed to higher levels of sulfur, effectively requiring lower levels of sulfur in the initial stream. Achievement of the 2015 target of tolerance to more than 100 ppm of sulfur will allow more flexibility in system design, especially if levels in the 1,000 ppm range can be achieved.

The membrane also must be structurally capable of withstanding the expected pressure drop across the system. Current coal gasification systems operate around 40 atmospheres of pressure; therefore, if the hydrogen product from the membrane is at 5–10 atmospheres, the differential pressure across the membrane would be 30–35 atmospheres (440–515 psi). Future coal gasification systems for hydrogen may operate at 80 atmospheres, so that the system pressure differential across the membrane could be as high as 70–75 atmospheres (1,000–1,100 psi). The membrane must also resist or be tolerant to atomic rearrangements, surface roughening, and formation of impurity over-layers that could adversely affect structural integrity in a WGS environment. In addition, it is critical that any membrane system be completely tolerant to carbon monoxide. It is also important to achieve higher hydrogen flux while simultaneously minimizing the pressure drop across the membrane in order to reduce the hydrogen product compression requirement. These target criteria are independent of membrane type.

4.4.2. Sorbent-based Systems Technical Targets

In sorbent-based processes, such as pressure swing adsorption or more advanced concepts such as chemical looping, the process performance depends on important parameters or characteristics for sorbents, which in turn affect energy balances, operating regimes or ranges, rates or throughputs of reactants, replacement rates for sorbents, and costs (capital and operating). In all sorbent-based processes, a sorbent material is exposed to the gas containing the components to be separated. Syngas from coal contains major species of H_2, CO_2, H_2O, and CO. For H_2 production, the CO needs to be "shifted" by reaction with H2O, producing more H_2 and CO_2. H_2 will be the product stream and needs to be produced at a pressure as close as possible to that needed for end use or transportation. The CO_2 needs to be produced at a pressure suitable for transportation to sequestration sites or other uses. Sorbent capacity is affected by all of the "state parameters" such as temperature and pressure, and will also depend on the partial pressures in the gas to which the sorbent is exposed. The sorbent will need to have a substantial adsorption capacity at the temperature planned for operation. The other critical parameters are the temperature and other conditions at which the products can be driven off the sorbent and the sorbent regenerated for use. If the temperature differential between adsorption and regeneration is too high, the regeneration heating requirements may be impractical.

Chemical looping takes advantage of the ability of some materials to adsorb and desorb oxygen in a cyclic manner, similar to the way that hemoglobin in the blood stream carries oxygen from the lungs to muscle cells. For gasification applications, the basic concept is to supply oxygen to the gasification process from a metal that is oxidized in air, and then chemically reduced with the fuel. Compared to conventional oxygen blown gasification, the process has an advantage in that no dedicated air-separation process is needed. CL can be used to produce hydrogen in gasification applications, where steam is used to oxidize the carrier, producing hydrogen. The metal oxide can again be recycled to a coal reactor where it would be reduced back to a metal.

During the last decade, the CL strategy has been applied to gasification to produce hydrogen from coal and coal derived syngas. The GE fuel-flexible process and the ALSTOM hybrid combustion– gasification process have been studied. These technologies use two different types of particles to convert coal into hydrogen: one type of particle is used to capture CO_2 while the other serves as an oxygen carrier.

More recently, the Ohio State University (OSU) has developed novel chemical looping gasification processes (i.e., the syngas chemical looping process, calcium looping process, and the coal/biomass direct chemical looping process). OSU is also developing a chemical looping scheme that could find application for treating tail gas from a coal based Fischer-Tropsch (F-T) Coal-to-Liquids process. This chemical looping concept uses iron oxide (Fe_2O_3) to react with the unreacted synthesis gas (H_2 and CO) and light hydrocarbons in the effluent tail gas from an F-T reactor. This reaction that takes place in a fuel reactor that produces CO_2, H_2O, and reduced iron. The reduced iron is then reacted with steam to produce hydrogen that can be recycled to the F-T reactor to adjust the input hydrogen to carbon monoxide ratio.

Lehigh University is developing the Thermal Swing Sorption Enhanced Reaction (TSSER) concept. One of the important applications of the TSSER concept is for gasification. The gas mixture from the gasifier would be fed into a TSSER system where the CO and H_2O would be catalytically converted to H_2 and CO_2 by the water gas shift reaction. The CO_2

would simultaneously be adsorbed from the reaction zone by a special chemisorbent material, resulting in a relatively pure product stream of CO_2-free H_2 gas. Removal of CO_2 from the reaction zone will drive the equilibrium controlled reaction to completion. This will result in extremely high conversion of CO and H_2O to H_2 in the sorber-reactor.

Specific milestones and targets for sorbent-based technologies are being developed in light of the Program's focus on electricity production.

4.4.3. Advanced Concepts Technical Targets

The basis for the 2015 technical targets assumes a single, compact WGS reactor operable over a wide range of temperatures and pressures with minimal undesirable side reactions and tolerance of common impurities found in coal-derived syngas. A catalyst lifetime of greater than 10 years is desirable, and depending on the form of the catalyst within the reactor, it may need to equal the expected operational life of the reactor. The cost goal is a 30 percent reduction over today's fixed-bed systems and a wider range of operating temperatures.

Partial oxidation of coal and other carbon-based solid/liquid feedstocks produces a synthesis gas with a composition ranging from 30–45 percent H_2, 35–55 percent CO, and 5–20 percent CO_2 (dry basis). This ratio can be adjusted to produce additional hydrogen. The WGS reaction converts CO and H_2O to CO_2 and H_2:

$$H_2O + CO \leftrightarrow 4\ H_2 + CO_2$$

This reaction also is used to increase the concentration of hydrogen in the syngas, and when coupled with an appropriate separation technology, it can produce high yields of high-purity hydrogen.

The WGS reaction is reversible, with the forward WGS reaction being mildly exothermic. Conversion to H_2 and CO_2 is thermodynamically limited and favored at lower temperatures. Higher temperatures improve the rate of reaction, but decrease the yield of hydrogen. In order to achieve high yields at high rates of reaction, the reaction is typically carried out in multiple adiabatic reactor stages, with lower reactor inlet temperatures in the latter stages. The yield also may be improved by using excess steam or by removing hydrogen to shift the WGS equilibrium to the right, in accordance with Le Chatelier's Principle. Steam also is used to minimize undesirable side reactions that compete with the WGS reaction. WGS catalysts and reactors could be improved by further R&D to increase hydrogen yield at higher operating temperatures, improve catalyst tolerance of syngas impurities, minimize undesirable side reactions, expand pressure and temperature operating ranges, and simplify/combine processing steps to reduce costs.

Some of the literature provides a considerable amount of experimental and modeling detail on WGSMRs, and there are several economic analyses (almost all on IGCC systems). In one study comparing chemical looping and WGSMRs, despite the lowest efficiency loss among the studied systems, the economic performance of the double stage chemical looping process was outperformed by WGSMRs and systems employing physical adsorption due to high costs. Processing modeling may be required to further study these advanced concepts, including co-production.

Specific milestones and targets for advanced concept technologies are being developed in light of the Program's focus on electricity production.

4.5. Technical Barriers

The following technical and economic barriers must be overcome to meet the goals and objectives of the Hydrogen Production Pathway for use in advanced IGCC power concepts.

4.5.1. General Barriers

A. High Cost. The cost of current technologies to produce hydrogen from coal must be reduced. This includes improved efficiency of the process, and reduced capital and operating costs.

B. Lack of Demonstration of Novel Technologies. Many novel separation processes (e.g., advanced membranes) have not been demonstrated at a scale sufficient to determine their potential for lower cost and efficient integration into advanced hydrogen from coal production systems.

C. Complex Process Designs. Complex process systems that have a greater number of process units require a larger plant footprint and are nearly always more difficult to improve in terms of efficiency. "Process intensification," in which multiple process function technologies are integrated into one process step — such as combined gas cleanup, WGS reaction, and hydrogen separation — offer potential advantages in scalability of the design, as well as better efficiency and lower costs. Various candidate process intensification processes and/or units require significant RD&D to establish their techno-economic viability.

4.5.2. Hydrogen Separation Barriers

There are several technology options available that can be used to separate hydrogen from synthesis gas. The following broad set of barriers must be overcome to reduce the cost and increase the efficiency of these separation technologies.

D. Loss of Membrane Structural Integrity and Performance. Depending on conditions, membranes may be subject to atomic rearrangements, surface roughening, pitting, and formation of impurity over-layers that may adversely affect structural integrity and performance. This becomes more important for the supported thin film membranes designed to enhance flux and minimize cost. For example, oxidizing gas mixtures (oxygen, steam, and carbon oxides) have been observed to cause metallic membranes to rearrange their atomic structure at temperatures greater than 450°C. This results in the formation of defects that reduce membrane selectivity for hydrogen. Some ceramic membranes exhibit poor thermo-chemical stability in CO_2 environments, resulting in the conversion of membrane materials into carbonates. In solvent systems, impurities can cause less effective absorption and may lead to excessive loss of solvent, which will increase cost and decrease separation efficiency.

E. Thermal Cycling. Thermal cycling can cause failure in some membranes, reducing durability and operating life.

F. Poisoning of Catalytic Surfaces. Metallic membranes must dissociate molecular hydrogen into hydrogen atoms before it can diffuse through the separation layer. The presence of trace contaminants, particularly sulfur, can poison the surface sites that are catalytically active for this purpose, diminishing the effectiveness of the membrane.

G. Defects During Fabrication. The chemical deposition of thin palladium or palladium-alloy membranes onto support structures is an important technical challenge in the fabrication of defect-free membranes. Large-scale, rapid manufacturing methods for defect-free thin films and membranes and modules in mass production must be developed and demonstrated. Fabrication of microporous membranes requires a reduction in membrane pore size, which is accomplished by deposition techniques. No synthesis and evaluation methods exist for tunable pore-size membranes used in separating H_2 from light gases at high temperature and in chemically challenging environments.

H. Lack of Seal Technology and Materials. High-temperature, high-pressure seals are difficult to make using ceramic substrates.

I. Technologies Do Not Operate at Optimal Process Temperatures. Membrane processes that can be designed to operate at or near system conditions, without the need for cooling and/or re-heating, will be more efficient.

J. Hydrogen Embrittlement of Metals. Below 300 °C, hydrogen can embrittle and induce a phase change in certain types of separation membranes. Embrittlement reduces the durability and effectiveness of the membrane for selectively separating hydrogen. Hydrogen also embrittles the structural steels of the membrane housing and gas handling systems.

K. Development of Lower Cost Non Precious Metal Hydrogen Separation Materials. Materials used in current hydrogen separation membranes are high in cost and not widely available from domestic sources.

4.5.3. Sorbent-based Systems Barriers

L. Low Adsorption Capacity. Current sorbent-based systems have limited adsorption capacity which increases the amount of sorbent required, increasing system capital and operating costs.

M. High Regeneration Energy Requirements. PSA and TSA systems require large variances in pressure and temperature to remove hydrogen. High heating, cooling, or pressurization requirements reduce efficiency and increase cost.

N. Technologies Do Not Operate at Optimal Process Temperatures. Sorbent-based processes that can be designed to operate at or near system conditions with minimized need for heating, cooling, or pressurization will be more efficient.

4.5.4. Advanced Concepts Barriers

O. Impurity Intolerance/Catalyst Durability. The WGS reaction occurs after coal has been gasified to produce synthesis gas. Impurities in the synthesis gas may act as poisons, deactivating the catalyst and damaging the structural integrity of the catalyst bed. Improved catalysts and reactor systems are needed to maintain catalyst activity throughout the reactor, and in some cases, eliminate the post-gasification synthesis gas cleanup step upstream of the WGS reactor.

P. Operating Limits. The synthesis gas produced from gasification exits the gasifier at a high temperature. The WGS reaction then is carried out in two separate stages: a high-temperature shift and a low-temperature shift. The development of advanced WGS catalysts and reactor systems that are more robust and can operate over a wide range of temperatures can eliminate the need for two separate stages, potentially reducing capital costs.

Q. Undesired Side Reactions. Reactions that produce species other than hydrogen and CO_2 must be minimized in the WGS reactor.

4.6. Technical Task Descriptions

Table 6 summarizes the tasks for the technologies under development.

Table 6. Task Descriptions for Hydrogen Production Technologies

Task Number	Task Description	Barriers Addressed by Task
1	*Advanced Hydrogen Separation* • Review and analyze separation technology to determine the current status, needs for advanced technology, preferred separation options, and scale-up to prepare modules. • Link membrane development work to material surface characterization studies in order to understand effects of impurities and operating conditions on short- and long-term membrane performance. • Conduct RD&D to explore technology for preferred advanced separation systems such as PSA, membranes, solvents, reverse selective systems, and other technology alternatives. • Identify low-cost materials, such as non-precious metals, for hydrogen separation. • Use molecular sieves to stabilize membranes. • Develop appropriate membrane seal and fabrication technologies and methods for module preparation and scale-up.	A, B, C, and D through K
2	*Sorbent-based Separation Systems* • Identify low-cost materials for CO2 separations. • Develop reverse selective hydrogen membranes for cost-effective separation of CO2 and other gases from mixed gas streams. • Develop advanced adsorption, hydrates, or other novel technologies for the cost-effective capture of CO2 from mixed gas streams.	L, M, N
3	*Advanced Concepts* • Investigate advanced and novel process concepts that integrate several processes — gas cleanup, WGS reaction, and hydrogen separation — into one step. • Investigate novel, "out-of-the-box" technologies that can produce hydrogen from coal directly or indirectly. • Develop advanced shift catalysts that are more active and are impurity-tolerant. • Conduct the WGS reaction using a high-temperature membrane without added catalyst. • Develop integrated single-step shift-membrane separation technology.	O, P, Q
4	*Demonstrations* • Demonstrate advanced hydrogen separation modules and technologies to confirm laboratory, bench-scale, and pre-engineering module results.	A, B

5. Implementation Plan

The Hydrogen from Coal Program was initiated in FY 2004 as a component of the overall DOE Hydrogen Program, and supports OCC's goals to develop technologies that would enable near-zero emissions coal facilities. The Program is in an operational mode, having initiated RD&D activities by requesting research proposals and selecting project performers. Continued execution and development of the Hydrogen from Coal Program requires proper management controls to ensure that the Program is progressing toward its goals and objectives.

5.1. Coordination with Other DOE/Federal Programs

The successful development of low-cost, affordable hydrogen production from fossil fuels coupled with sequestration of CO_2 is dependent on technologies being developed in a number of ongoing associated RD&D programs within FE and NETL. These technologies are needed for:

- CO_2 capture and sequestration.
- Advanced coal gasification, including feed handling systems.
- Efficient gasifier design and materials engineering.
- Advanced synthesis gas cleanup technologies.
- Advanced membrane separation technology to produce a lower-cost source of oxygen from air.
- Fuel cell modules that can produce electric power at coal-fired integrated gasification combined-cycle power plants.
- Hydrogen fuel gas turbines.

In response to comments by the National Academy of Sciences, the Hydrogen from Coal Program was organizationally grouped together with the Carbon Sequestration Program to enhance coordination and collaboration with respect to carbon sequestration and hydrogen production from coal. Figure 11 shows the various programs and projects with which the Hydrogen from Coal Program will coordinate in addition to the Sequestration Program. Coordination of efforts and sharing of information and experience will help ensure the successful transition to a hydrogen energy system.

5.1.1. Other Coordination Activities

The Hydrogen from Coal Program interacts with several different programs and federal organizations outside of FE. These include the overall DOE Hydrogen Program, the Hydrogen Interagency Task Force, and the International Partnership for a Hydrogen Economy (IPHE).

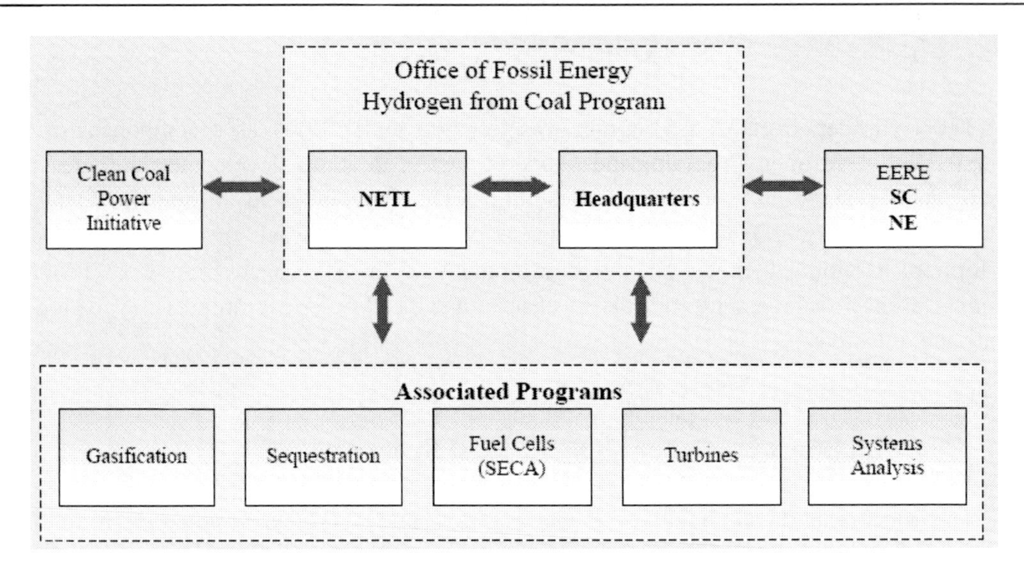

Figure 11. Coordination of the Hydrogen from Coal Program with Associated Programs.

5.1.1.1. Hydrogen and Fuel Cell Interagency Task Force

The Hydrogen and Fuel Cell Interagency Task Force was established in April 2003 to fulfill a statutory requirement and to serve as a mechanism to facilitate collaboration among federal agencies engaged in hydrogen and fuel cell R&D activities. In November 2006, the Hydrogen and Fuel Cell Technical Advisory Committee (HTAC), an advisory committee to the Secretary of Energy, recommended that agency members on the Interagency Task Force be represented at the Assistant Secretary level to ensure a continued high-level of commitment and decision-making on hydrogen activities. The Assistant Secretary for Fossil Energy represents the Hydrogen from Coal Program on the Interagency Task Force. The Task Force, chaired by the Assistant Secretary of EERE, held its initial meeting in August 2007. The Task Force also includes representatives from the following organizations:

- DOE Offices of Energy Efficiency and Renewable Energy; FE, Nuclear Energy, and Science
- National Institute of Standards and Technology (NIST)
- Department of Defense (DoD)
- Department of Transportation (DOT)
- Department of Education
- Department of Agriculture (USDA)
- Executive Office of the President, Office of Science and Technology Policy (OSTP)
- General Services Administration (GSA)
- United States Postal Service (USPS)
- Environmental Protection Agency
- National Aeronautics and Space Administration
- National Science Foundation (NSF)

5.1.1.2. International Partnership for the Hydrogen Economy (IPHE)

The IPHE was established in 2003 and consists of 16 countries and the European Union (EU). The Hydrogen from Coal Program contributes to the IPHE by attending meetings and offering its expertise on hydrogen from coal production technologies. The partners of the IPHE include nearly 3.5 billion people; account for more than $35 trillion in Gross Domestic Product (GDP) (approximately 85 percent of the world's GDP), and more than 75 percent of electricity used worldwide; and produce greater than two-thirds of CO_2 emissions, while consuming two-thirds of the world's energy. The IPHE focuses its efforts on:

- Developing common codes and standards for hydrogen fuel utilization.
- Establishing cooperative efforts to advance the RD&D of hydrogen production, storage, transport, and end-use technologies.
- Strengthening exchanges of pre-competitive information necessary to build the kind of common hydrogen infrastructures necessary to allow this transformation to take place.
- Formalizing joint cooperation on hydrogen R&D to enable sharing of information necessary to develop hydrogen-fueling infrastructure.

5.2. Performance Assessment and Peer Reviews

Performance assessment provides essential feedback on the effectiveness of the Program's mission, goals, and strategies. It is built into every aspect of program management and provides managers with a consistent stream of information upon which to base decisions about program directions and priorities. The overall DOE Hydrogen Program has annual merit review meetings of funded projects to report progress and provide program managers the opportunity to evaluate progress toward program goals and milestones. Additionally, NETL hosts periodic peer reviews that are conducted by the American Society of Mechanical Engineers (ASME) to evaluate progress and provide guidance and direction.

The RD&D Plan will be annually reviewed and updated to reflect Department priorities, changes in technical and economic assumptions, and accomplishments of its research activities. These annual reviews of the RD&D Plan will provide program managers the opportunity to update the goals and objectives of the Program by utilizing the most current data generated by the Program and consistent with DOE and NETL's OCC management guidance. On a periodic basis, program managers will provide RD&D direction and the project managers will conduct reviews to evaluate progress toward goals. The project managers will provide their input into the RD&D Plan by review and comment on individual projects, and their assessment of the progress being made to achieve the program goals, milestones, and targets. Formal meetings will be held with the NETL Technology Manager and DOE Program Manager on an annual basis and fact sheets will be provided on individual projects on a periodic basis.

5.3. Accomplishments and Progress

The Hydrogen from Coal Program has successfully transitioned from its initial start-up in FY 2004 to full operations. The Program has been actively soliciting proposals from industry, universities, and other organizations to help achieve its goals. Currently, the program has 23 projects that are conducting research in a wide number of areas (Table 7).

Table 7. Active Hydrogen from Coal Research Projects

Research Area[a]	Number of Projects
Palladium and metallic-based membrane research	6
Module scale-up	5
Membrane reactors & process intensification	1
Microporous membrane research	1
Sorbent/chemical looping	3
Non-precious metal separations	7
TOTAL	*23*

[a] Complementary projects are supported by the Gasification and Sequestration Programs.

5.3.1. Technical Progress

The Hydrogen from Coal Program has been in existence since 2004, with most of its projects initiated in FY2005 and thereafter. Since its inception, the Program has made significant technical progress toward achieving several of its goals, milestones, and technical targets. Several of the activities undertaken by the program have produced advancements and progress in technology development as outlined in the next several paragraphs.

5.3.1.1. Hydrogen Membranes

Several of the hydrogen membrane developers have obtained laboratory results that indicate their membranes can achieve the Program's flux Technical Targets using pure gases without sulfur, and some test units show resistance to carbon monoxide and hydrogen sulfide in syngas. NETL has developed and published a Hydrogen Membranes Test Protocol for use by all contractors, and NETL's Office of R&D (ORD) has conducted verification testing.

- Eltron Research, Inc. is developing alloy-based membranes and has developed a separator unit rated to produce 1.5 lbs/day of hydrogen. Eltron's best alloy membrane has demonstrated a H_2 flux rate of 411 SCFH/ft^2 at specified pressure and gas compositions, which is higher than the Program's Technical Target. Eltron initiated membrane tests under WGS feed stream conditions; tubular membranes were successfully tested for greater than 300 hours with feed gas composition of 50 percent H_2, 29 percent CO_2, 19 percent H_2O, 1 percent CO, and 1 percent He. A lifetime testing reactor has been operated to 600 hours; initial baseline membrane testing in H_2/N_2 feed streams show stable membrane hydrogen flux performance at

200 SCFH/ft^2. A preferred membrane coating catalyst was tested in streams with 20 ppm H_2S, and stable H_2 flux was observed for 160 hours.

- Worcester Polytechnic Institute (WPI) is developing Pd-based membranes on tubular stainless steel or alloy supports. WPI achieved a H_2 flux of 359 SCFH/ft^2 in pure gases at 442°C and 100 psi ΔP with a 3–5 µm thick Pd membrane with an Inconel base, exceeding the DOE flux target. WPI has also built an engineering-scale prototype membrane with 8.8 µm thickness, 2" outer diameter, and 6" length. WPI has demonstrated long term membrane testing with total test duration of 63 days at 450°C, 15 psi ΔP, 80 SCFH/ft^2 H_2 flux, 99.99 percent purity. This flux is equivalent to approximately 340 SCFH/ft^2 under DOE flux target operating pressures. WPI also is developing new membranes utilizing SS 316L supports.
- Praxair is building a Pd-alloy based prototype multi-tube hydrogen purifier based on U.S. Patent 7,628,842, and will use the unit to demonstrate prototype performance. A Pd-Au membrane (5 percent Au, 9µm thickness) was prepared from an extruded substrate and tested at 400°C at a pressure range of 20–200 psi. The H_2 flux in pure gas was 384 SCFH/ft^2 and the H_2/N_2 selectivity was 495 at 200 psi. Another Pd-Au membrane (9 percent Au, 8µm thickness) prepared from an extruded tube, tested at the same conditions, had an infinite selectivity up to 100 psi and a higher H_2 flux, most likely due to a thinner Pd layer.
- United Technologies Research Center (UTRC) is developing two types of membrane separators using metallic supports: one based on a commercial tube design with Pd-based alloy; the other using a novel nano-composite material. UTRC has tested five separators using PdCuTM alloy which showed increased surface stability in bench-scale tests. Sulfur tolerance has been tested at up to 78 ppm H_2S. UTRC's current membrane flux performance is approximately 45 SCFH/ft^2 at a temperature of 450°C and feed pressure of 200 psi, below the DOE Target. In addition to the Pd-based metallic membranes, the project is also focused on developing an advanced membrane concept which is a hybrid of ceramic and dense metallic membrane technology. Anticipated performance for the nano-oxide membrane projects H_2 flux of 400 SCFH/ft^2.
- SwRI is developing Pd-foil-based membrane separators. SwRI has tested 70 percent Pd, 20 percent Au, 10 percent Pt foils under the DOE Test Protocol gas composition, with flux levels of approximately 30 SCFH/ft^2. Testing was conducted done at 400°C, 170 psi with gas composition of 50 percent H_2, 30 percent CO_2, 19 percent H_2O, and 1 percent CO.

Some of the recent contractor results are summarized in Figure 12 showing their relationship to the Program Technical Targets for hydrogen separation membranes.

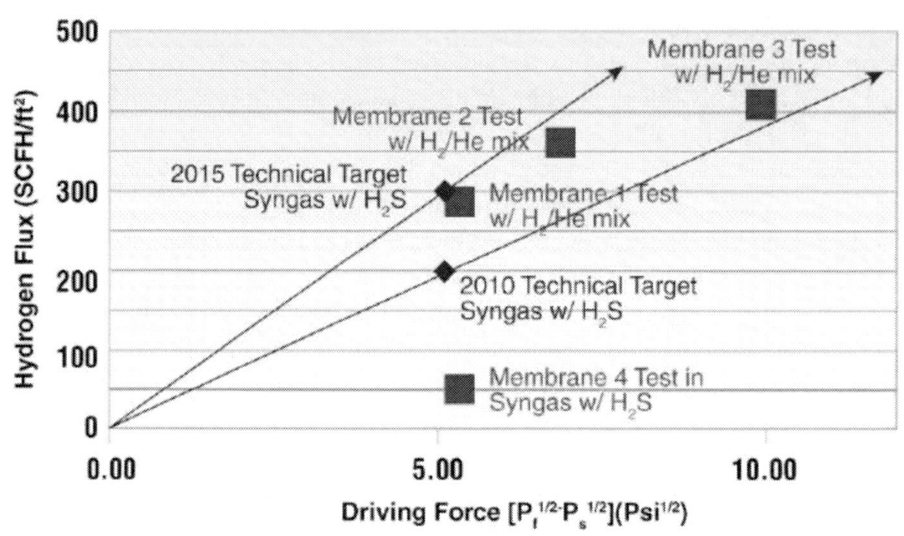

Note: 2010 and 2015 Technical Targets extrapolated over all driving force ranges based on Seivert's Law.

Figure 12. Membrane Flux Measurements from Research Projects Sponsored by the Hydrogen from Coal Program.

In the advanced concepts area of process intensification, a WGS membrane reactor is being developed by several organizations, including Media and Process Technologies. Their goal is to meet performance requirements in terms of H_2/CO selectivity (50 to less than 100), hydrothermal stability (50 psia steam), and chemical stability (resistance to sulfur and hydrocarbons poison). A field test at a commercial hydrotreating facility was conducted that successfully demonstrated selectivity and chemical stability in a gas stream containing H_2, hydrocarbons, H_2S and ammonia.

NETL's in-house research group also conducts its own exploratory research in the areas of membranes and catalysis in support of the coal-based hydrogen production pathway. Researchers have performed independent verification testing of several membranes. They have also studied the impacts of sulfur on palladium-type membranes and have shown two degradation mechanisms occur and that the concentration of gas species throughout the membrane reactor can have a critical role on membrane degradation.

5.3.1.2. Systems Engineering

Systems engineering analytical activities support the Hydrogen from Coal program by providing: a) current cost estimates for hydrogen production and associated plant/component configurations; and b) R&D guidance regarding the materials, equipment, and system configurations that are likely to offer optimum efficiency and cost. The results of these studies provide assessments of the current and future state of the technologies under development, and help guide program direction.

NETL completed a comprehensive assessment of the production of precious metals currently used in hydrogen membrane construction. The assessment showed:

- Global deposits are primarily located in South Africa and Russia.

- Less than 10 significant mining companies currently exist in the world; production is declining.
- Commercial deployment using precious metals has potential global economic and environmental impacts.

Each of these factors could restrain the ability to deliver hydrogen via membrane separation technologies. Considering these concerns, a competitive Funding Opportunity Announcement (FOA) was released soliciting research projects that would conduct both fundamental and applied research on novel non-precious metal hydrogen separation concepts. The projects were selected in September 2009 and are shown in Table ES-1.

5.4. Communications, Outreach, and Technology Transfer

Information dissemination, communications, and outreach activities are important and integral parts of the Hydrogen from Coal Program. Program officials communicate the Program's mission, strategies, accomplishments, and technology capabilities to a variety of stakeholder audiences including Congress, the public, educational institutions, industry, and other government and non-government organizations. Program staff perform the following communications, outreach, and technology transfer in addition to their other programmatic duties:

- Present technical status and program overviews at public forums.
- Manage the FE and NETL public Web site, and document and references lists.
- Manage official correspondence.
- Coordinate reviews of FE/NETL-related statements by other DOE offices and federal agencies.

The Program also participates in various technical conferences and workshops to exchange information with industry, government, and academia throughout the world.

5.5. Next Steps

The Hydrogen from Coal Program has transitioned from its FY2004 initial start-up mode to an operational/implementation mode. The Program will continue to issue solicitations as appropriate, and will continue with current RD&D activities that support development and deployment of hydrogen from coal technologies to address the overall DOE Hydrogen activity goals of improved energy security and reduced GHG emissions.

NETL has implemented a peer review process that provides for input from technical experts in academia, industry, and other stakeholder organizations on the Strategic Center for Coal's programs.

The Hydrogen from Coal Program has undergone one such review and another is scheduled for October 2010. The recommendations from these reviews are addressed from program and technical perspectives to improve the overall quality of the Program and help ensure that goals and targets are met successfully and on schedule. It is intended that this process will continue as the technologies mature.

As previously identified as part of the RD&D Plan review, FE/NETL management has directed the Program to focus on production of hydrogen for use in stationary turbines and possibly solid oxide fuel cells (SOFC) to produce clean electricity from coal. The RD&D plan will continue to be updated periodically based on RD&D progress and subsequent go/no-go decisions and funding appropriations. Systems analysis and evaluation will continue to guide the direction of research and provide input into the Hydrogen from Coal RD&D Plan. The Program will continue collaborating with associated programs in OCC to ensure efficient utilization of resources and successful development and integration of hydrogen from coal technologies into clean coal processes. The Program also will continue to work closely with EERE, SC, and NE on coordinating activities within DOE to meet its goals and objectives.

6. Appendix A – Additional Technology Discussion

6.1. The Relationship between Membrane Flux and Partial Pressure of Hydrogen

With some exceptions, most hydrogen separation membrane research is in the laboratory research phase. Therefore, most of the current information on membranes, particularly the flux, is based on observed data under specifically controlled experiments that may not reflect real-world operating conditions in a hydrogen from coal production facility. However, based on scientific and engineering theory and observation data in the laboratory, estimates of the hydrogen flux at desired operating pressures can be determined.

As previously mentioned, membrane flux is dependent upon the partial pressure of hydrogen, and the relationship between the two differs depending upon the type of membrane. Specifically, microporous membranes exhibit a flux that is directly proportional to the hydrogen partial pressure differential across the membrane. In metal or hydrogen-permeable cermet membranes, the flux is proportional to the difference in the square roots of the partial pressures or the natural log of the partial pressure gradient according to Sievert's Law. In dense ceramic and non-hydrogen permeable cermets, flux is proportional to the natural log of the pressure gradient across the membrane, based on the Nernst potential.

Flux rates need to be converted from observed experimental results to desired operating pressure conditions to evaluate their status relative to technical targets. Table 8 shows these mathematical relationships for the different membrane types.

Table 8. Relationships for Flux as a Function of Hydrogen Partial Pressure Differentials for Different Membrane Types

Membrane Type	ΔP function	Equation
Microporous	Linear	$Flux_{est}$ M = $Flux_{obs}$ M* (ΔP_{est}/ ΔP_{obs})
Pure metallic (includes pure metal and metal alloys)	Square root	$Flux_{est}$ P = $Flux_{obs}$ P* $[(Pf_{est}^{0.5} - Ps_{est}^{0.5})/(Pf_{obs}^{0.5} - Ps_{obs}^{0.5})]$
Hydrogen-permeable cermet	Square root	$Flux_{est}$ P = $Flux_{obs}$ P* $[(Pf_{est}^{0.5} - Ps_{est}^{0.5})/(Pf_{obs}^{0.5} - Ps_{obs}^{0.5})]$
Dense ceramic	Natural logarithm	$Flux_{est}$ D = $Flux_{obs}$ D*$[\ln(Pf_{est}/Ps_{est})/\ln(Pf_{obs}/Ps_{obs})]$
Dense ceramic with non-hydrogen permeable second phase (electron conducting)	Natural logarithm	$Flux_{est}$ D = $Flux_{obs}$ D*$[\ln(Pf_{est}/Ps_{est})/\ln(Pf_{obs}/Ps_{obs})]$

Fluxest M is the estimated flux for microporous membranes.
Fluxobs M is the observed, or tested, flux for microporous membranes.
ΔP est is the ΔP of hydrogen partial pressure to be estimated.
ΔP obs is the observed, or tested, hydrogen partial pressure.
Fluxest P is the estimated flux for hydrogen permeable metallic, metal alloy, or cermet membranes.
Fluxobs P is the observed, or tested, flux for hydrogen permeable metallic, metal alloy, or cermet membranes. Pfest is the estimated feed side hydrogen partial pressure.
Psest is the estimated sweep (permeate) side hydrogen partial pressure.
Pfobs is the observed, or tested, feed side hydrogen partial pressure.
Psobs is the observed, or tested, sweep (permeate) side hydrogen partial pressure.
Fluxest D is the estimated flux for dense ceramic or non-hydrogen permeable cermet membranes.
Fluxobs D is the observed, or tested, flux for dense ceramic or non-hydrogen permeable cermet membranes.

Figure 13 shows the effect of changes in partial pressure on the flux of hydrogen membranes. This graph is based on a reference assumed flux of 60 SCFH/ft2 with a hydrogen partial pressure ΔP of 20 psi and an assumed sweep (permeate) side hydrogen partial pressure of 1 psi for all membrane types. For commercial applications, the sweep, or permeate, side hydrogen partial pressure is assumed to be 50 psi.

One of the key conclusions observed from Figure 13 is that it is important to set desired technical targets near the expected operating conditions. In the case of hydrogen from coal technologies, hydrogen separation membranes are expected to operate with at least 50 psi hydrogen partial pressure on the permeate side and a hydrogen partial pressure ΔP of 100–300 psi is expected. For example, when converting assumed observed test data from a ΔP of 20 psi and a permeate side partial pressure of 1 psi to operating conditions of 100 psi ΔP and 50 psi permeate side, a decline in flux for dense ceramic membranes is seen, a slight increase for Pd-type, but a linear improvement related to ΔP for microporous membranes.

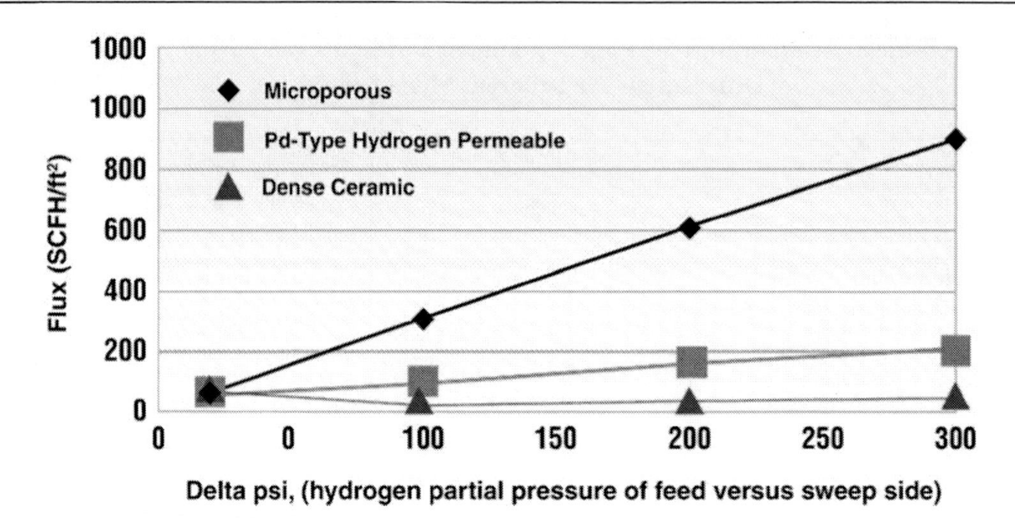

Figure 13. Ideal Effect of Changes in ΔP on Flux of Hydrogen Membranes.

6.2. Sorbents

Carbon Dioxide Sorbents

For testing and screening of sorbents, the most basic dataset is an absorption isotherm for single gases for a single temperature at a time over a range of temperatures at various pressures (e.g., amount of CO_2 adsorbed per unit weight of sorbent plotted versus partial pressure). At a minimum, all sorbent candidates should have adsorption isotherms determined for CO_2 and H_2 in the temperature and pressure ranges listed below. Selectivities can be calculated from single gas adsorption isotherms, but these calculations may not accurately represent some interactive effects (e.g., catalysis of the shift reaction). Global kinetics can be calculated from the time dependence of the TGA data, but single purpose kinetics equipment may be needed to provide more accurate information. Thermal properties of the sorbent such as heat capacity can be determined using calorimeters, and these properties may vary with temperature.

In general, sorbents intended for use for separation of CO_2 from syngas need to be tested in the following ranges:

- Temperature – 150 to 500°C (e.g., ~ 150–200°C for sorption and 250°C for regeneration)
- Pressure – 40 to 60 barg (580 to 870 psig)
- Gas composition – Typical syngas compositions

Once adsorption isotherms and other related data are available over the temperature and pressure ranges of interest, process models can be used to estimate the overall performance of the sorbent in a CO_2 capture process located at a given position in the overall H_2 production process. Process models will need to simulate the energy balance (energy in and out), as well as flows and compositions in and out of the process. An energy balance calculation will need to calculate: (1) energy flows during adsorption which is expected to be exothermic (i.e., the sorbent will be heated up during adsorption and heat may need to be removed via heat exchangers in the bed), and (2) energy flows during additional heating of and/or gas flows through the sorbent needed to cause release of the adsorbed CO_2 and regeneration of the sorbent. Energy flows will be affected by heat capacities and heat transfer rates of the vessels used to house the sorbent as well as sorbent properties. It may be possible to get reasonable estimates by ignoring vessel heating, but heat removed or added by heat exchangers cannot be ignored.

Results of screening tests and process calculations can be used to "weed out" sorbents that are not promising because of low capacity or unfavorable kinetics or selectivity. Sorbents that pass the screening tests can move on to be tested for process-specific applications.

Hydrogen Sorbents

As indicated in the main text, in considering whether to use a hydrogen sorbent in a TSA or PSA process, the economics of the process will depend (among other things) on the ratio of H_2 to CO_2 in the syngas, the compression requirements following separation of the two main streams, and how well the sorbent discriminates between H_2 and CO_2. PSA processes usually produce a higher pressure stream from the gases that pass over the sorbent and a lower pressure stream from the gases that are adsorbed and later desorbed. Since coal contains far more carbon by weight than hydrogen, on a superficial basis it would seem that more energy would be required in recompressing CO_2 than H_2, and therefore it would be preferable to allow the H_2 to be produced at a lower pressure than the CO_2. However, the shift reaction creates more H_2 from CO in the syngas, and one has to look at the overall ratio in the shifted syngas. Typical shifted syngas from coal gasification may contain 40–50 percent H_2 and 40–50 percent CO_2 on a molar basis, but on a weight basis far more CO_2 is produced. It is necessary to examine in detail how the volume, weight and other properties of the gases affect compression requirements.

PSA reactors are operated in a cyclic process. For PSA reactors in which the sorbent adsorbs CO_2 and other non-hydrogen gases, assuming a freshly regenerated sorbent, the major steps are: pressurization, feed, blowdown, purge, and pressure equalization.

- The first step is to pressurize the bed with the input gas. The feed stage involves pumping the mixed gas through the sorbent bed under high pressure, with the bed prepared in advance by covering with hydrogen. This step allows CO_2 and other non-H_2 gases such as CO and water to be adsorbed, and pure hydrogen to pass through.
- When the sorbent bed is nearly saturated with CO_2 and other gases, some of the pressure is released and the remaining hydrogen is allowed to flow out at a lower pressure.
- Then the flow is reversed, allowing the bed to blowdown to a low pressure (typically atmospheric) that releases the adsorbed gases.
- A purge flow is then created with hydrogen or recycled product gas to cause the remaining adsorbed gases to be released due to a reduction in partial pressure of the adsorbed gases.
- The bed is then partially pressurized with the feed gas and then the pressurization started again with product gas to ready the bed for feed again.

In the case of an H_2 sorbent, the sorbent will either react chemically with H_2 or adsorb it by a physical adsorption process. How much H_2 can be adsorbed by the sorbent (sorbent capacity) is characterized by the percentage weight increase of the sorbent when the sorbent is exposed to the mixed gas containing H_2 under fixed conditions of temperature, pressure, gas composition, and flow rate. The sorbent capacity is affected by all of the "state parameters" such as temperature and pressure, and will also depend on the partial pressure of H_2 in the gas to which the sorbent is exposed. The sorbent will need to have a substantial adsorption capacity at the temperature planned for operation. The speed of H_2 removal will be affected by the kinetic rates of reaction. Ideally, the sorbent will adsorb only H_2 and all the CO_2, CO, etc. will remain in the gas phase. In practice, it may be necessary to measure the selectivity to H_2 over other gas components. The other critical parameter is the temperature and other conditions at which the H_2 can be driven off the sorbent and the sorbent regenerated for use. If the temperature differential between adsorption and regeneration is too high, the regeneration heating requirements may be impractical.

6.3. WGS Technologies

One or two staged reactors are typically employed in commercial WGS technology to produce hydrogen by steam reforming of natural gas. Commercial catalysts have been developed to achieve optimum performance in the different stages and are summarized in Table 9. Only fixed-bed reactors are currently used in commercial applications with these catalysts. Multiple reactors with inter-cooling are used to optimize the WGS reaction temperature profile. Steam reforming plants typically employ either a two-stage system using high (Fe/Cr) and low (Cu/Zn) temperature shift catalysts in series, or a single stage with high- or medium-temperature shift catalyst followed by a PSA hydrogen separation system. Partial oxidation plants used to gasify oils, coke, and coal employ multiple reactor stages using either the high-temperature or sour gas (Co/Mo) shift catalyst in all beds. No gas cleanup is required upstream of the WGS reactors with the sour gas shift catalyst. For low-temperature shift, catalyst life is limited due to loss of activity. For high-temperature shift, catalyst life is limited due to increases in pressure drop and loss of activity. Technology options for residual CO cleanup/H_2 purification include methanation (old), PSA (current), and polymer membranes (new). Possible impurities in the product hydrogen are CO, CO_2, CH_4 and higher hydrocarbons, and methanol.

In summary, the advantages of low-/medium-temperature shift processes are:

- WGS equilibrium favors hydrogen production at low temperatures, maximizing hydrogen yield.
- Undesirable side reactions like F-T synthesis are minimized.
- Processes integrate well with conventional gas cleanup technologies that produce hydrogen at near-ambient temperatures and pipeline pressures (400 psi); minimal or no reheat required.
- Temperature range overlaps ranges for advanced gas cleanup processes for sulfur, mercury, etc.
- Processes can be coupled with newer preferential oxidation (PrOx) technologies to produce very low CO in the hydrogen product.
- Steam requirements are low.

Table 9. Performance of Commercial WGS Catalysts

Performance Criteria	Units	Low-/Medium-temperature Shift	High-temperature Shift	Sour Gas Shift
Catalyst form	–	Pellets	Pellets	Pellets
Active metals	–	Cu/Zn & Cu/Zn/Al	Fe/Cr	Co/Mo
Reactor type	–	Multiple fixed beds (last bed)	Multiple fixed beds	Multiple fixed beds
Temperature[a]	°C	200–270/300	300–500	250–550
Pressure	psia	~450	450–750	~1,100
CO in feed	–	Low	Moderate to high	High
Residual CO	%	0.1–0.3	3.2–8	0.8–1.6
Approach to	°C	8–10	8–10	8–10
Min steam/CO ratio	Molar	2.6	2.8	2.8
Sulfur tolerance	ppmv	<0.1	<100	>100[b]
COS conversion	–	No	No	Yes
Chloride tolerance	–	Low	Moderate	Moderate
Stability/durability	Years	3–5	5–7	2–7

[a] Lower temperature limit is set by water dew point at pressure.
[b] Sulfur is required in the feed gas to maintain catalyst activity.

The disadvantages are:

- WGS kinetics are more favorable at higher temperatures.
- Low-temperature shift catalysts are easily poisoned.
- Temperature range is below the range of metal and ceramic membranes that could be used for separation.
- Copper (Cu) in catalyst promotes methanol side reaction (methanol emissions from hydrogen plants are regulated by the U.S. Environmental Protection Agency).
- Any condensation of water in the reactor will irreversibly damage the catalyst. The advantages of high-temperature shift processes are:
- WGS kinetics improve with higher temperatures.
- Processes can operate at very high pressures (~1,000 psi).
- Catalysts exhibit greater tolerance for potential poisons.
- Temperature range is consistent with metal and ceramic membranes. The disadvantages are:
- WGS equilibrium is less favorable at higher temperatures.
- Undesirable side reactions (F-T synthesis) are favored at higher temperatures.
- Steam requirement increases with temperature, both to improve equilibrium and minimize side reactions.
- Hexavalent chromium (from the catalyst) presents a wastewater treatment and catalyst disposal issue.

7. APPENDIX B

7.1. Acronyms

Government Agency/Office Acronyms

DoD	Department of Defense
DOE	Department of Energy
DOT	Department of Transportation
EERE	Office of Energy Efficiency and Renewable Energy
EPA	Environmental Protection Agency
FE	Office of Fossil Energy
GSA	General Services Administration
NASA	National Aeronautics and Space Administration
NE	Office of Nuclear Energy
NETL	National Energy Technology Laboratory
NIST	National Institute of Standards and Technology
NSF	National Science Foundation
OCC	Office of Clean Coal
ORD	NETL's Office of Research & Development
OSTP	Executive Office of the President – Office of Science and Technology Policy
SC	Office of Science
USDA	United States Department of Agriculture
USPS	United States Postal Service

General Acronyms

ΔP	Delta P (change in pressure)
°C	degrees Celsius
°F	degrees Fahrenheit
CCPI	Clean Coal Power Initiative
CCS	Carbon Capture and Storage
CL	Chemical Looping
cm^2	Square centimeter
COE	Cost of electricity
EPRI	Electric Power Research Institute
EU	European Union
FOA	Funding opportunity announcement
ft^2	Square feet
ft^3	Cubic feet
F-T	Fischer-Tropsch
FY	Fiscal year
GDP	Gross domestic product
GHG	Greenhouse gas
GTI	Gas Technology Institute
HEV	Hybrid electric vehicle
HRSG	Heat recovery steam generator

IGCC	Integrated Gasification Combined Cycle
IPHE	International Partnership for the Hydrogen Economy
K	Degrees Kelvin
kg	Kilogram
kWh	Kilowatt-hour
LHV	Lower heating value
min	Minute
ml	Milliliter
MMBtu	Million British thermal units
μm	Micrometer
mill	One tenth of one cent
MW	Megawatts
MWh	Megawatt-hour
N/A	Not available
N/D	Not demonstrated
nm	Nanometer
OSU	Ohio State University
PC	Pulverized coal
PM	Particulate matter
ppb	Parts per billion
ppbv	Parts per billion on a volume basis
ppm	Parts per million
ppmv	Parts per million on a volume basis
ppmw	Parts per million on a weight basis
PrOx	Preferential oxidation
PSA	Pressure swing adsorption
psi	Pounds per square inch
psia	Pounds per square inch absolute
psig	Pounds per square inch gauge
R&D	Research and development
RD&D	Research, development, and demonstration
SCFH	Standard cubic feet per hour
SCR	Selective catalytic reduction
SNG	Substitute natural gas
SOFC	Solid oxide fuel cell
SwRI	Southwest Research Institute
TSA	Thermal swing adsorption
TSSER	Thermal Swing Sorption Enhanced Reaction
U.S.	United States
UTRC	United Technologies Research Center
WGS	Water-gas shift
WGSMR	Water-gas shift Membrane Reactor
WGCU	Warm gas cleanup
WPI	Worcester Polytechnic Institute

Chemical Symbols/Names

Ag Silver
Al Aluminum
Ar Arsenic
Au Gold
Ba Barium
Br Bromine
Ca Calcium
CH_4 Methane
Cl Chlorine
CO Carbon monoxide
CO_2 Carbon dioxide
COS Carbonyl sulfide
Cr Chromium
Cu Copper
F Fluorine
Fe Iron
Fe_3O_4 Synthetic Iron Oxide (Magnetite or Iron Oxide Black)
H_2 Hydrogen
Hg Mercury
H_2O Water
H_2S Hydrogen sulfide
HClH ydrogen chloride (hydrochloric acid)
HCN Hydrogen cyanide
Mg Magnesium
N_2 Nitrogen
NH_3 Ammonia
Ni Nickel
NO_x Nitrogen oxides
O_2 Oxygen
Pb Lead
Pd Palladium
SO_2 Sulfur dioxide
V Vanadium
Zr Zirconium

End Notes

1 DOE/NETL, Affordable, Low-Carbon Diesel Fuel from Domestic Coal and Biomass, January 2009
[2] DOE/NETL, Current and Future Technologies for Gasification-Based Power Generation, November 2009
[3] DOE/NETL, Cost and Performance Baseline for Fossil Energy Plants, May 2007
[4] DOE/NETL, NETL Test Protocol: Testing of Hydrogen Separation Membranes, October 2008
[5] Impact of CO_2 Capture on Transport Gasifier IGCC Power Plant, Bonsu, A., et. al., Southern Company Services – Power Systems Development Facility; Booras, G., Electric Power Research Institute (EPRI), Breault, R., NETL; Salazar, N., Kellogg, Brown and Root, Inc., International Technical Conference on Coal Utilization and Fuel Systems, Clearwater, FL, May 21–25, 2006.

[6] Comparison of a New Warm-Gas Desulfurization Process versus Traditional Scrubbers for a Commercial IGCC Power Plant, presented at Gasification Technologies Conference, Oct 17, 2007 by Jerry Schlather and Brian Turk (RTI/Eastman Chemical).

In: Clean Energy Solutions from Coal
Editor: Marcus A. Pelt
ISBN: 978-1-61324-724-2

Chapter 3

NETL TEST PROTOCOL: TESTING OF HYDROGEN SEPARATION MEMBRANES*

National Energy Technology Laboratory

DISCLAIMER

This report was prepared as an account of work sponsored by an agency of the United States Government. Neither the United States Government nor any agency thereof, nor any of their employees, makes any warranty, express or implied, or assumes any legal liability or responsibility for the accuracy, completeness, or usefulness of any information, apparatus, product, or process disclosed, or represents that its use would not infringe privately owned rights. Reference therein to any specific commercial product, process, or service by trade name, trademark, manufacturer, or otherwise does not necessarily constitute or imply its endorsement, recommendation, or favoring by the United States Government or any agency thereof. The views and opinions of authors expressed therein do not necessarily state or reflect those of the United States Government or any agency thereof.

NETL Contact:
Daniel Driscoll
Fuels Division
Strategic Center for Coal

National Energy Technology Laboratory
www.netl.doe.gov

* This is an edited, reformatted and augmented version of a National Energy Technology Laboratory publication, DOE/NETL – 2008/1335, from www.netl.doe.gov, dated October 2008.

Prepared by:

Daniel Driscoll
Fuels Division
Strategic Center for Coal

Brian Morreale
Office of Research and Development

Supported by:
Larry Headley
TMS, Inc.
Contract Number
DE-AC26-05NT41816

Acknowledgments

The authors would like to acknowledge and thank Daniel Cicero for the concept of this protocol, and the other members of the NETL Hydrogen and Syngas Technology Team for comments and contributions. This report does not contain any sensitive or proprietary information.

Objective

The overall objective of this testing is to develop membrane technologies that achieve target performance of hydrogen separation membranes for use in a gasification system process. NETL needs (1) to know whether developers are approaching/achieving Hydrogen Program Technical Targets, (2) to be able to compare results on an "apples to apples" basis, and (3) to clearly state expectations to contractors. Results may be of value in deciding on down-selections. Another objective is to determine the suitability of each membrane and to assess the compatibility of each membrane's optimum operating conditions for use in plant hardware as it currently exists.

This test protocol provides background information on gasification systems, descriptions of membrane tests, the prescribed procedures for tests, and the analyses and reporting to be completed. Very briefly, a series of tests is described beginning with no gasification contaminants to determine ideal membrane performance, followed by tests with sulfur, corresponding to sulfur levels for membranes aimed at different process configurations. Should any membrane not perform satisfactorily during the first two tests, the membrane shall be considered to be disqualified from further testing at larger scales and/or more stringent conditions. Future tests will be considered at higher contaminant levels and/or raw syngas streams.

BACKGROUND

Advancements in hydrogen membrane separation technologies have the potential to reduce costs, improve efficiency, and simplify hydrogen production systems. Desirable characteristics of separation membranes are high hydrogen flux at low pressure drops; tolerance to contaminants, including sulfur species, CO, NH_3, chlorides, As, and Hg; mechanical strength; low cost; and operation at a range of system temperatures to provide versatility in location of membrane separators in the sequence of process operations. Many current hydrogen membrane technologies are at the research stage, but some have the potential to provide hydrogen purity above 99.99 %.

The DOE Hydrogen from Coal RD&D Plan[1] establishes Technical Targets for Hydrogen Separation Membranes that are necessary to meet the performance and cost goals of the program. The reader is referred to the RD&D Plan for rationale and more detailed descriptions of the Targets. Note: For the convenience of readers, this document will quote temperatures in both °C and °F, pressures in common engineering units of psia, psig, and atmospheres, fluxes in standard cubic centimeters per minute per square centimeter of membrane area ($sccm/cm^2$) and standard cubic feet per hour per square foot of membrane area ($SCFH/ft^2$), and gas compositions in molar percent and/or parts-per-million (ppm). The Technical Targets are as follows:

- 2010 Target
  - Hydrogen flux – 200 $SCFH/ft^2$ (~ 100 $sccm/cm^2$) @ 100 psi ΔP H_2 partial pressure. Standard conditions are 150 psia hydrogen feed pressure and 50 psia hydrogen sweep pressure.
  - Temperature – 300 to 600 °C (572 to 1112 °F).
  - Pressure performance – ΔP – Up to 400 psi.
  - Sulfur tolerance – 20 ppm.
  - CO tolerance – Yes.
  - Water-gas-shift (WGS) activity – Yes.
  - Hydrogen purity – 99.5 %.
- 2015 Target
  - Hydrogen flux – 300 $SCFH/ft^2$ (~ 150 $sccm/cm^2$) @ 100 psi ΔP H_2 partial pressure. Standard conditions are 150 psia hydrogen feed pressure and 50 psia hydrogen sweep pressure.
  - Temperature – 250 to 500 °C (482 to 932 °F).
  - Pressure performance – ΔP – 800 to 1000 psi.
  - Sulfur tolerance – >100 ppm.
  - CO tolerance – Yes.
  - WGS activity – Yes.
  - Hydrogen purity – 99.99 %.

Gasifier Process Conditions and Syngas Compositions

Hydrogen separation membranes may be used in a variety of locations in a gasification-based coal-to-hydrogen production process, depending on the capability of the membrane to withstand temperature and pressure conditions as well as variations in gas composition. The following discussion briefly outlines some of those configurations. The coal gasifiers currently under consideration for coal-to-hydrogen or co-production plants that produce electricity and hydrogen are listed in Table 1, along with typical temperature and pressure conditions and gas compositions at a point in the process downstream of the first stages of gas cooling and particulate cleanup. Figure 1 shows the main process units that the gas stream encounters prior to this point.

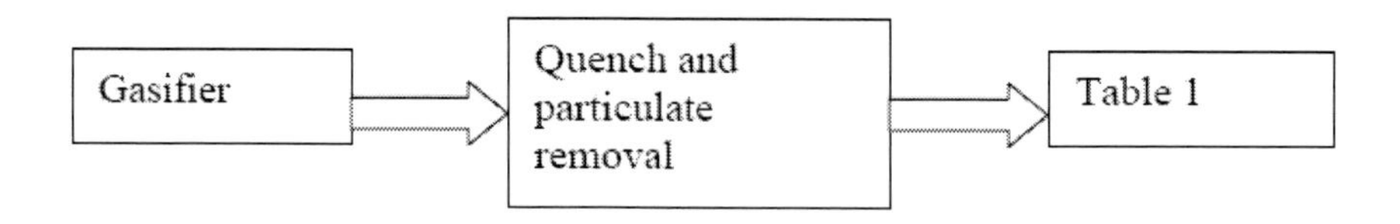

Figure 1. Block Diagram of Syngas Process Units Corresponding to Table 1.

The primary contaminants of the syngas are sulfur (H_2S and COS) and nitrogen compounds (NH_3), but there also may be small amounts of chlorides (such as HCl) and mercury. Total sulfur concentrations (typically most of the sulfur is H_2S) in gasifier raw gases are usually in the 1% range. Ammonia (NH_3) generally is in the 0.1 to 0.2% range. Mercury concentrations are difficult to measure, but are estimated to be approximately 50 parts-per-million by weight (ppbw) based on mercury content in most coals[2]. Hydrogen chloride (HCl) concentrations may range from 50 to 500 ppm[3], and these are a concern regarding corrosion in heat exchangers. The pressure and temperature conditions here would represent the harshest conditions to which hydrogen separation membranes would likely be exposed if used in conjunction with advanced shift reactor processes in the raw gas stream.

Table 1. Gasifier Syngas Conditions and Compositions Upstream of Shift Reactors

Vendor/Gasifier	GE Energy[4]		Conoco-Phillips[5]		KBR Transport[6]		Shell[7]	
Syngas Pressure (psia)	800		614		400+		565	
Syngas Temperature [°C (°F)]	210 (410)		927 (1700)		260-371 (500-700)		260 (500)	
Syngas Composition (mole %)	H_2	25.9	H_2	26	H_2	29.2	H_2	15.6
	CO	26.7	CO	37	CO	34.3	CO	30.7
	CO_2	11.6	CO_2	14	CO_2	13.6	CO_2	1.13
	H_2O	33.6	H_2O	15	H_2O	18.9	H_2O	48.3
	CH_4	0.08	CH4	4	CH_4	2.5	CH_4	0.02
	H_2S	0.56	H_2S	0.51	H_2S	0.056	H2S	0.43
	COS	0.01	COS	0.00	COS	?	COS	0.04
	NH_3	0.13	NH_3	0.19	NH_3	0.28	NH_3	0.18
					HCN	0.032		

The membrane must be structurally capable of withstanding the expected pressure drop across the system. In general, current coal gasification systems operate around 40 atmospheres of pressure; therefore if the hydrogen product from the membrane is at 5–10 atmospheres, the differential pressure across the membrane would be about 30–35 atmospheres or 450–525 psi. Future coal gasification systems for hydrogen may operate at 80 atmospheres, so that the system pressure differential across the membrane could be as high as 800–1,000 psi. The Program Targets require membranes to withstand a pressure differential of 400 psi (2010 Target) or 800- 1000 psi (2015 Target). Corresponding to the total pressure conditions and hydrogen contents in the tables above, implied maximum available hydrogen partial pressures in these systems would be as follows:

- GE – 334 psi.
- Shell – 266 psi.
- E-Gas 212 psi.

According to DOE's H_2A analysis protocol[8], the desired hydrogen delivery pressure to the plant gate of a central hydrogen production plant is 300 psi. Clearly, considering additional membrane separation pressure drops, in some cases additional end-stage or inter-stage compression may be necessary to achieve this delivery pressure. Process designs for hydrogen plants might try to maximize gasifier pressure to achieve high ΔP and hydrogen flux, and/or minimize the need to recompress hydrogen. However, higher pressure gasifiers produce more methane, so a tradeoff study might be necessary to optimize pressure selection. In any case, the feed pressures and maximum pressures listed in the Program Targets above are reasonably consistent with available hydrogen partial pressures for gasification systems.

Water-Gas-Shift (WGS) Reactor Process Conditions

Shift reactors must be used to convert water and CO in the gas into additional H_2 and CO_2 in order to maximize the H_2 content for separation and make CO_2 available for sequestration. Raw gas shift reactors using sulfur resistant catalysts can be applied to the gas before desulfurization. Alternatively, clean synthesis gas after sulfur removal can be reheated to 315-371 °C (600-700 °F) for WGS reactors that produce additional hydrogen through the catalytic reaction of CO with H_2O to form CO_2 and H_2.

As mentioned earlier, there are several possible strategies for placement of hydrogen separation membranes in a coal gasification process. One strategy is to position the membrane downstream of the shift reactors; another is to place membrane separators between shift reactors, while a third is to integrate shift reaction with hydrogen separation to use removal of hydrogen from the syngas as a means to drive the shift reaction to completion at higher temperatures than would normally be utilized in a shift reactor. The latter innovative design is referred to as a water-gas-shift membrane reactor (WGSMR). Because the WGS reaction is exothermic and a large amount of shift is expected to occur within the membrane reactor, the membrane reactor also should operate in a temperature range compatible for the WGS to occur. WGS activity is an essential function of the membrane reactor for coal-to-hydrogen applications. In addition, for metallic membranes where catalytic activity for hydrogen dissociation is important, tolerance to sulfur compounds such as H_2S and COS is desirable. Failure to achieve sulfur tolerance would require an additional sulfur polishing step in the coal-to-hydrogen plant configuration.

The following information on shift reactor conditions is taken from the DOE Hydrogen from Coal RD&D Plan1. Typically shift reactors may be designed as multiple stages, with the first stages operating at higher average temperatures to take advantage of better kinetics, and later stages operating at lower average temperatures because of equilibrium chemistry that favors higher hydrogen contents (driving the shift reaction to the right). Because of the exothermic nature of the shift reaction, temperature will increase at the reactor exit unless internal cooling is provided. The gas is cooled between stages by heat exchange and/or steam addition. Typical reactor conditions and catalyst properties for the various types of shift reactors are as follows:

- Raw or sour gas (RG) shift
  - Temperature range – 250 to 550 °C (482 to 1022 °F).
  - Catalyst – Co/Mo (the catalyst is actually a sulfided form of Co and Mo).
  - Sulfur tolerance – > 100 ppmv (sulfur is required for the catalyst).
- High temperature (HT) shift
  - Temperature range – 300 to 500 °C (572 to 932 °F).
  - Catalyst – Fe/Cr.
  - Sulfur tolerance – < 100 ppmv .
- Low Temperature (LT) shift
  - Temperature range – 200 to 300 °C (392 to 572 °F).
  - Catalyst – Cu/Zn or Cu/Zn/Al.
  - Sulfur tolerance – < 1 ppmv.

The NETL IGCC studies discussed above use water gas shift before cold desulfurization, which implies that raw gas shift is used in these systems. The Eastman Chemical/RTI warm gas cleanup system has desulfurization before water gas shift, which implies that HT shift is used.

Since hydrogen separation membranes may alternatively be installed in the gasification system process stream following shift reactors, gas conditions and hydrogen content following shift reactors are listed in Table 2. (This information is taken from reference 4.) Figure 2 shows the main process units that produce these compositions.

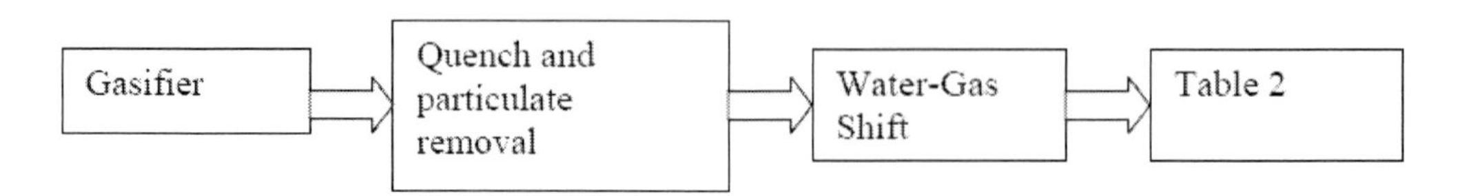

Figure 2. Block Diagram of Syngas Process Units Corresponding to Table 2.

The primary changes from Table 1 are that 112 and CO_2 levels are higher and CO levels are significantly lower. Steam levels are considered to be variable. All cases use raw or sour gas shift, so sulfur and nitrogen species will be similar to the values in Table 1. This report does not consider the TRIG gasifier, but these compositions would be expected to be similar.

Table 2. Gasifier Syngas Conditions and Compositions Downstream of Shift Reactors

Vendor/Gasifier	GE Energy Radiant	Conoco- Phillips E-Gas	Shell
Syngas Pressure (psia)	777	516	483
Syngas Temperature [°C (°F)]	270 (519)	236 (457)	35 (95)[9]
Syngas Composition	112 43%	112 41%	112 55%
(only major gases	CO 0.9	CO 0.5	CO 1.7
are shown)	CO_2 31	CO_2 32	CO_2 38
(mole %)	H_2O 24	H_2O 22	H_2O 0.1

Warm Gas Cleanup Process Conditions

Warm gas cleanup[10] can improve the efficiency of the gasification process by eliminating the need to cool the gas all the way to ambient temperature before separation and/or utilization. A typical process sequence design for warm gas cleanup in a hydrogen production process is particulate removal, gas cleanup, shift reactor, and hydrogen separation. Figure 3 shows the main process units in this system.

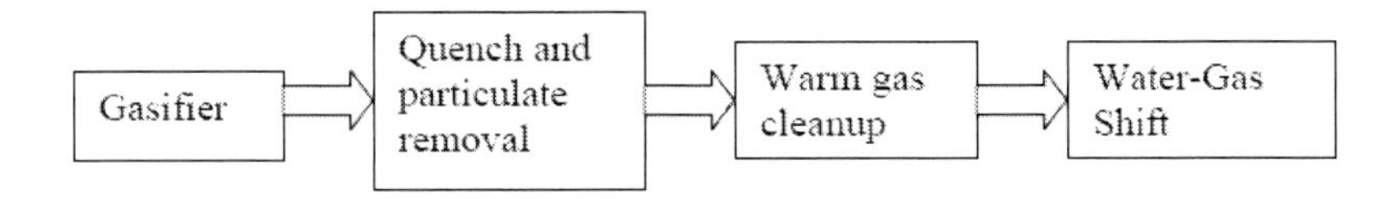

Figure 3. Block Diagram of Warm Gas Cleanup System.

Detailed gas compositions are not available for this system. However, typical warm gas desulfurization conditions are as follows:

- Desulfurization temperature range – 315 to 538 °C (600 to 1000 °F).
- Desulfurization pressure range – 300 to 600 psig.
- Inlet sulfur concentration – 7000 to 8661 ppmv (parts-per-million-by-volume).
- Outlet sulfur concentration – 0.4 to 20.6 ppmv.

For warm gas contaminant removal (other than sulfur), these conditions generally apply:

- Temperature range – > 250 °C (482 °F).
- Ammonia – regenerable – removal to < 50 ppmv.
- Mercury, other – disposable sorbents.

Cold Gas Cleanup

With current commercial syngas cleaning technology, the gas has to be cooled to ambient temperature to remove contaminants. Depending on the system design, a series of scrubbing processes are used to remove particulates, H_2S, COS, and NH_3, and many of these systems operate at low temperatures with synthesis gas leaving the process at about room temperature. The H_2S and COS, once hydrolyzed, are removed by dissolution in, or reaction with, an organic solvent and converted to valuable by-products, such as elemental sulfur or sulfuric acid with 99.8% sulfur recovery. Cold gas desulfurization systems applied to coal gasification can reduce H_2S levels to approximately the parts-per-million (ppm) range. In a design study using the Selexol process,[11] syngas is cleaned to less than 1 ppm total sulfur, and the study states that the sulfur level can be further reduced to less than 1 ppb using a zinc oxide sulfur polishing bed.

Potential Sulfur Effects on Metal-based Membranes

Metallic membranes must dissociate molecular hydrogen into hydrogen atoms before diffusion through the separation layer. The presence of trace contaminants, particularly sulfur, can poison the surface sites that are catalytically active for this purpose, diminishing the effectiveness of the membrane. The shift reaction and hydrogen extraction by means of a membrane separation system will have a major influence on the driving force for poisoning or corrosion in the membrane. Figure 4 shows a simulation of this effect that was conducted by NETL's Office of Research and Development using the Comsol scientific simulation package.[12] The curves correspond to mole fractions of H_2, CO, CO_2, H_2O, and H_2S in the syngas as it flows in a Water- Gas-Shift Membrane Reactor (WGSMR) that has a cylindrical shape. The general gas flows along the axis of the reactor while the hydrogen (permeate) flows radially. The effect of hydrogen being removed by the membrane as the gas flows axially is to drive the shift reaction towards completion, reducing the CO and H_2O levels, while increasing CO_2, and also increasing the mole fraction of H_2S in the gas, since H_2S is not removed with H_2. The driving force for membrane corrosion is the ratio of H_2S to H_2 in the surface region. Since H_2 is being removed, and the H_2S mole fraction is increasing, the H_2S/H_2 ratio will increase considerably by the time the gas mixture reaches the downstream end of the reactor. Thus levels of H_2S that are noncorrosive at the reactor inlet may become extremely corrosive at the reactor outlet, which could influence the chemical stability of the membrane. This is also the case for oxidation, which is driven by the ratio of CO to CO_2. Membrane testing must be done with cognizance of these issues.

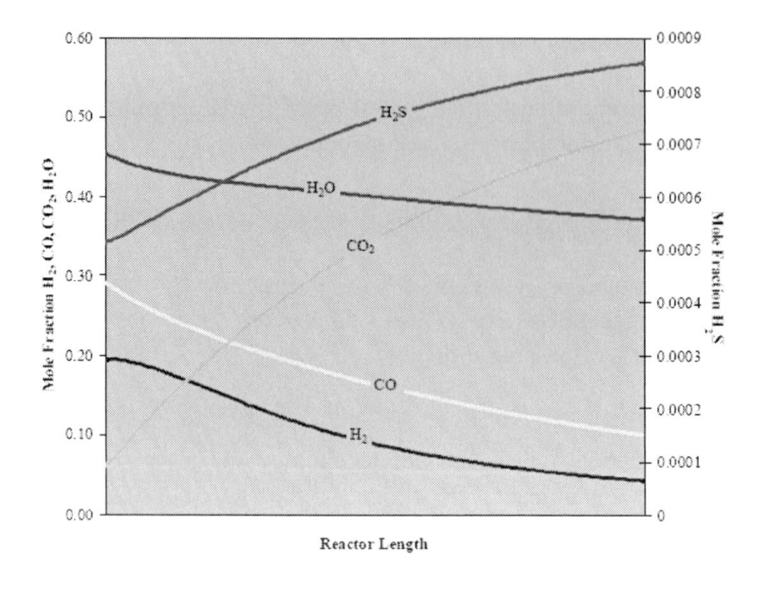

Figure 4. Comsol® simulated gas composition of a syngas mixture along the length of the Water Gas Shift Membrane Reactor (WGSMR) (permeance used is comparable to a 25 micron Pd-foil).

TESTING PROTOCOL DESCRIPTION

General Comments on Testing

The objective of the testing is to qualify the membrane for further consideration in larger scale performance tests of multiple cycles. The membrane formulation and configuration in the bench reactor should be evaluated primarily by:

- Hydrogen separation efficiency and hydrogen flux of the membrane during testing in constant multi-gas loading (calculated at the gas inlet and outlet locations to the membrane unit).
- Measurement of flux and separation factor as a function of gas composition and contaminant level.
- Mechanical and physical properties of the fresh and exposed membranes (e.g., morphology changes and chemical degradation resistance).

For each test, NETL will specify:

- Temperature of membrane and feed gas.
- Feed gas and sweep gas composition.
- Feed and sweep pressure.
- Time of each phase of test.
- Data to be collected during test and data collection frequency.
- Post-test analysis.
- Review of test plan and data analysis.

Membrane Characteristics

The membranes to be tested are mixed metal, metal-ceramic, or pure metal or ceramic formulations which may be thin film or have either metal or ceramic support material. The morphology may vary and may have been produced through various methods. It is anticipated that most of the membranes to be tested in the near term are likely to be constructed of metals that transport hydrogen in the atomic form and generally follow the Sievert's Law pressure dependence exhibited by palladium or palladium-alloy membranes.

Summary of Test Phases

The two near-term test phases will be based on increasing "reality" of gas composition and mechanical stress, and on increasing stringency of hydrogen flux performance. For example, the first test will be with no sulfur or other contaminants, and the first sulfur exposure test will be for sulfur levels comparable to warm gas cleanup. Later tests will use increasing sulfur levels all the way to raw gas. This outline provides a brief description of the purpose and rationale for the tests; additional detailed descriptions are provided in the next section.

- Test 1 – 2010 Technical Target Screening Test without Sulfur Matches sulfur content for cold gas cleanup with guard bed and LT WGS
- Test 2 – 2010 Technical Target Qualification Tests Matches temperature and sulfur content for warm gas cleanup and HT WGS
- Future Tests – Testing with partially cleaned gas or raw gas from gasifier

Since gas cleanup, WGS, and membrane separation may be closely coupled or integrated in coal-based hydrogen production system designs, it is important to understand these relationships and accommodate them in the testing protocol. Table 3 summarizes the compatibility of hydrogen separation membranes that successfully pass the test phases with the various types of gas cleanup and WGS reactors that might be encountered in a coal gasification-to-hydrogen plant. .

Table 3. Membrane Compatibility with Process Conditions if Successful in Testing

	Membrane Compatibility with Coal-to-Hydrogen Plant Process Conditions		
Test Phase	Syngas Cleanup Processes	Water Gas Shift (WGS) Processes	General
Test 1	• Requires low temperature desulfurization and possibly zinc oxide guard bed for zero sulfur level	• LT WGS	• Tolerant to syngas CO • Could be placed after LT WGS or staged with LT WGS • Contaminant effects unknown
Test 2	• Low temperature desulfurization • Warm gas cleanup that achieves < 20 ppm H_2S	• LT WGS	• Could be placed after LT WGS or staged with LT • WGS Allows tubular membrane operation with 20 ppm H2S • Other contaminant effects unknown
Future Tests	• Partially cleaned gas • Raw gas before desulfurization	• HT WGS • Raw gas WGS	• Suitable for staged WGS or WGS Reactor

TEST SPECIFICATIONS

Test 1 – 2010 Technical Target Screening Test Without Sulfur

The objective of this test stage is to determine whether the proposed membrane design, composition, and construction are capable of meeting the Technical Targets of the program with the exception of resistance to contaminants such as sulfur. The lack of sulfur compounds in the stream is potentially achievable by cold gas cleanup with a zinc oxide guard bed and LT WGS.

The gas composition specified is similar to that produced by the coal gasifiers described above with the exception of contaminant species. The conditions generally correspond to hydrogen separation after cold gas desulfurization including zinc guard bed for final polishing, and LT shift. See Table 4 for detailed specifications of gas composition, feed and sweep pressures, and temperatures. The test will be conducted under the following conditions:

- Temperature – Selected by contractor in range of 300 to 600 °C (572 to 1112 °F).
- Feed gas composition – Syngas after shift using simulated gas (bottled gases).
  - 5–10% He will be added to the feed gas for leak detection and the compositions in Table 4 adjusted to keep the same ratios of other gases.
- Sulfur content – Zero.
- Feed hydrogen differential partial pressure relative to sweep gas– ~100 psi.
- Sweep gas – Tests can be run without sweep, but if sweep gas is used it is recommended that argon be used.

Please note that for these tests

- A leak test will be conducted on each membrane to be tested.
- The target for hydrogen purity is 99.5%.
- Hydrogen flux will be determined at above conditions at constant temperature and flow conditions over an 8 hour period, as a minimum, and reported as an hourly average.
- It is recommended that each membrane be tested to 120 hours to determine whether there are long term performance or structure changes.

Test 2 – 2010 Technical Target Qualification Tests

The objective of this series of tests is to determine whether the targeted membrane hydrogen separation efficiency and flux can be maintained with at least 99.5% H_2 purity in the presence of sulfur contamination up to the 2010 Technical Target level. This series of tests will also determine the potential effects of increasing H_2S/H_2 ratio due to removal of H_2 through the membrane as the feed gas passes along the membrane from inlet to outlet.

Baseline Test

Each test phase in Test 2 will begin with a new membrane. Each membrane used in Test 2 will be subjected to a baseline test at the conditions specified in Test 1 to ensure that its performance is reasonably consistent with the original membrane tested in Test 1. The duration of the baseline test will be decided by the contractor as long as it is sufficient to make a flux determination and determine that the membrane is not leaking. Flux with each new membrane should be reproducible within plus or minus 20% of the flux achieved during the original Test 1.

Test Conditions

The tests in Test 2 will be conducted at the conditions specified below. These conditions generally correspond to those that would be encountered by hydrogen separation membranes installed downstream of warm gas desulfurization and HT shift. See Table 4 for detailed specifications of gas composition, feed and sweep pressures, and temperatures for the following three tests (2A, 2B, and 2C):

- Temperature – Selected by manufacturer in range of 300 to 600 °C (572 to 1112 °F).
- Sulfur (H_2S) content in feed gas – 20 ppm.
- Feed hydrogen differential partial pressure relative to sweep gas – ~100 psi.
- Sweep gas – Tests can be run without sweep, but if sweep gas is used it is recommended that Argon be used.

Please note that for these tests

- A leak test will be conducted on each membrane to be tested.
- The target for hydrogen purity is 99.5%.
- Hydrogen flux will be determined at constant temperature and flow conditions over an 8 hour period, and reported as an hourly average.
- It is recommended that each membrane be tested to 120 hours to determine whether there are long term performance or structure changes.

- Conditions for Test 2A:
  - Feed gas composition – Syngas after shift.
  - 5–10% He will be added to the feed gas for leak detection and the compositions in Table 4 adjusted keeping the same ratios of other gases.
- Conditions for Test 2B:
  - Feed gas composition – Syngas after 50% of H_2 has been removed from gas specified in Test 2A.
  - 5–10% He will be added to the feed gas for leak detection and the compositions in Table 4 adjusted keeping the same ratios of other gases.
- Conditions for Test 2C:
  - Feed gas composition – Syngas after 95% of H_2 has been removed from gas specified in Test 2A.
  - 5–10% He will be added to the feed gas for leak detection and the compositions in Table 4 adjusted keeping the same ratios of other gases.

Detailed Test Conditions

Table 4 below details the test conditions for Tests 1 and 2.

Table 4. Summary of Test Conditions for Tests 1 and 2

	Test Phase			
	Test 1	**Test 2A**	**Test 2B**	**Test 2C**
Feed Gas Composition (Before correction for addition of He)				
H_2 (%)	50.0	50.0	33.0	4.8
CO (%)	1.0	1.0	1.3	2.0
CO_2 (%)	30.0	30.0	40.0	57.0
H_2O (%)	19.0	19.0	25.0	36.2
H_2S (%)	0.000	0.002	0.003	0.004
Sweep Gas Inlet Composition				
Ar (%)	100	100	100	100
Total Feed Pressure (psia)	200	200	200	200
H_2 Feed Partial Pressure (psia)	100	100	66	9.6
Total Sweep Pressure (psia)	<30	<30	<30	<30
(PH_2S/PH_2)Feed	0.00E+00	4.00E-05	9.09E-05	8.33E-04
Temperature (^oC)	300-600	300-600	300-600	300-600

Future Tests with Partially Cleaned or Raw Syngas

NETL is contemplating future tests to determine hydrogen membrane separation efficiency and flux in the presence of sulfur contamination up to and above the 2015 Technical Target level, extending to raw gas conditions of sulfur levels and other gasifier syngas contaminants. These tests will be further defined as the earlier testing (Tests 1 and 2) are completed. Generally this testing will be intended to address conditions corresponding to use of a hydrogen separation membrane in a Water Gas Shift Reactor configuration in partially cleaned or raw gas from one of the selected gasifiers. The test protocol for these tests will be specified by the NETL COR.

- Feed gas composition would be selected corresponding to syngas after quench and particulate removal but before shift reactors or other gas cleanup systems.
- Sulfur content and other contaminants would be equivalent to those in gasifier raw gas.
- Total pressure would be in the 500 – 800 psi range or possibly higher, depending on the status of gasifier development and the results of systems studies for hydrogen production.
- Exposure to actual gasifier raw gas is desirable if sidestream test facilities are available.

Data Collection and Analysis

The following measurements will be made and reported as a minimum for each test condition. Note - Standard temperature and pressure (STP) = 60 °F (15.6 °C), 14.7 psia.

- Membrane temperature (°C).
- Time required for heat-up and conditioning.
- Membrane area and thickness (cm^2 and cm).
- Retentate Pressure (kPa and psia).
- Retentate Flow reported at STP.
- Retentate composition (before and after reactor).
- Sweep Pressure (kPa and psia).
- Sweep Flow reported at STP.
- Sweep composition (before and after reactor).
- Measurements of gas composition by gas chromatograph.

Please note that

- Gas compositions and flows will be reported as hourly averages for an 8 hour test period and for long-term tests (120 hours), if conducted.
- Flows will be reported in sccm, temperatures in °C, pressures in kPa and psia.

Analyses will include the following and will be described in detail sufficient to allow the reader to recalculate the results from experimental data:

- Hydrogen Flux at STP for test conditions – hourly averages for the 8 hour test period. Hydrogen flux will be reported both in $sccm/cm^2$ and in $SCFH/ft^2$ of membrane area.
- Calculations of hydrogen fluxes at STP for "standard" conditions of 150 psi feed hydrogen partial pressure and 50 psi hydrogen sweep partial pressure for each of the hourly averages.
- Calculations of hydrogen purity.

Reporting

A comprehensive test topical report shall be completed and provided to DOE within two weeks of the completion of each test. All experimental data and calculations will be made accessible to NETL personnel for review.

The minimal outline of this report shall include:

1) Introduction.
2) Test Protocol Modifications (if any).
3) Experimental Procedures.
4) Test Results for Separation.
5) Post-Test Characterization.
6) Summary and Conclusions.

End Notes

[1] Hydrogen from Coal RD&D Plan, External Draft, Sept. 2007

[2] The Cost of Mercury Removal in an IGCC Plant, NETL Final Report, Sept. 2002, Parsons Corporation.

[3] High Temperature Corrosion in Gasifiers, Bakker, W., Materials Research, 2004, v. 7, No. 1, pp. 53-59.

[4] Data from NETL Cost and Performance Baseline for Fossil Energy Plants, Volume 1:Bituminous Coal and Natural Gas to Electricity Final Report (Bituminous Baseline [BB] Report), DOE/NETL-2007/1281, May 2007, Exhibit 3- 33, Stream 9 upstream of shift reactor.

[5] Pressure, temperature, and major gas species (H_2-CH_4) from NETL Gasification Overview Presentation, July 2007. Sulfur and ammonia values from Cost and Performance Baseline for Fossil Energy Plants, Volume 1:Bituminous Coal and Natural Gas to Electricity Final Report, DOE/NETL-2007/1281, May 2007, Exhibit 3-66, Stream 9 after shift reactors. Note – BB report does not show stream composition before shift reactor.

[6] Data from Power Systems Development Facility Update on Six TRIGTM Studies, presented at the International Pittsburgh Coal Conference 2006.

[7] Data from NETL Cost and Performance Baseline for Fossil Energy Plants, Volume 1:Bituminous Coal and Natural Gas to Electricity Final Report (BB Report), DOE/NETL-2007/1281, May 2007, Exhibit 3-99, Stream 13 upstream of water scrubber.

[8] http://www.hydrogen

[9] Shift and additional gas cooling are combined in this system.

[10] Comparison of a New Warm-Gas Desulfurization Process versus Traditional Scrubbers for a Commercial IGCC Power Plant, presented at Gasification Technologies Conference, Oct 17, 2007 by Jerry Schlather and Brian Turk (RTI/Eastman Chemical).

[11] Baseline Technical and Economic Assessment of a Commercial Scale Fischer-Tropsch Liquids Facility DOE/NETL-2007/1260.

[12] Bryan Morreale, NETL ORD Hydrogen Separation Group Leader, private communication, December, 2007.

In: Clean Energy Solutions from Coal
Editor: Marcus A. Pelt
ISBN: 978-1-61324-724-2

Chapter 4

COAL-TO-LIQUIDS TECHNOLOGY: CLEAN LIQUID FUELS FROM COAL*

United States Department of Energy

COAL — NOT AN ORDINARY ROCK

Coal is a solid fossil fuel with a high carbon content but a low hydrogen content, typically no more than 5–6 percent of the total weight of coal. On a molecular level, it consists of long chains of mostly aromatic hydrocarbon structures. It is mostly associated with the generation of electric power or as a feedstock in the production of steel. However, this versatile, solid rock can be broken down into simple molecules and put back together into many diff erent, useful forms.

Technologies exist today to break down coal into the simple molecules of carbon monoxide and hydrogen, and then to combine these molecules to form many useful products such as liquid transportation fuels, natural gas, and chemical feedstocks that are used to produce common household products such as tape and fi lm. Today, these products are manufactured using fuels and chemicals produced from petroleum and natural gas. However, the United States has the opportunity to more fully utilize its abundant coal resource as a fl exible feedstock to produce liquid fuels and chemicals that address the country's energy and economic needs through Coal-to-Liquids (CTL) technology.

The United States has an abundance of coal — approximately a 250-year supply at today's production rates.

* This is an edited, reformatted and augmented version of the United States Department of Energy publication, dated March 2008.

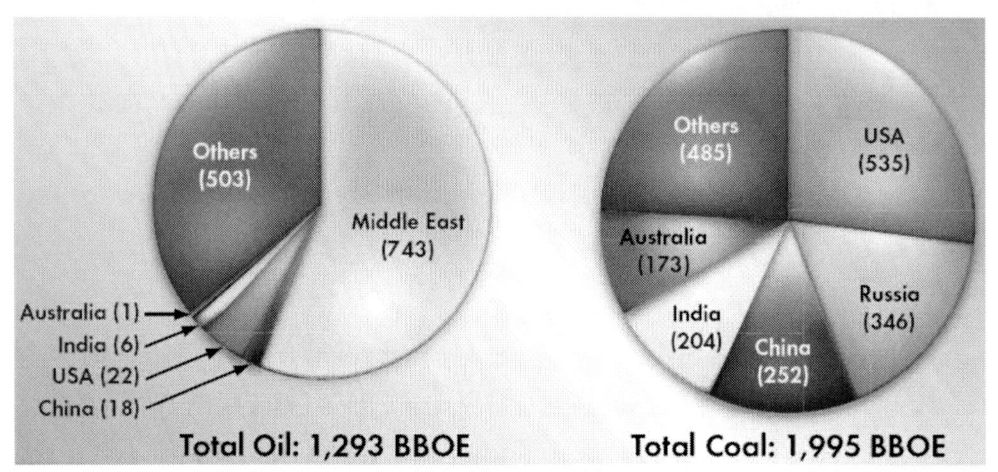

* Source: EIA, International Energy Annual 2005 (1 ton coal is equivalent to 2 BBOE) and Oil and Gas Journal, January 200.

Figure 1. World Oil and Coal Reserves — Billion Barrels Oil Equivalent (BBOE)*.

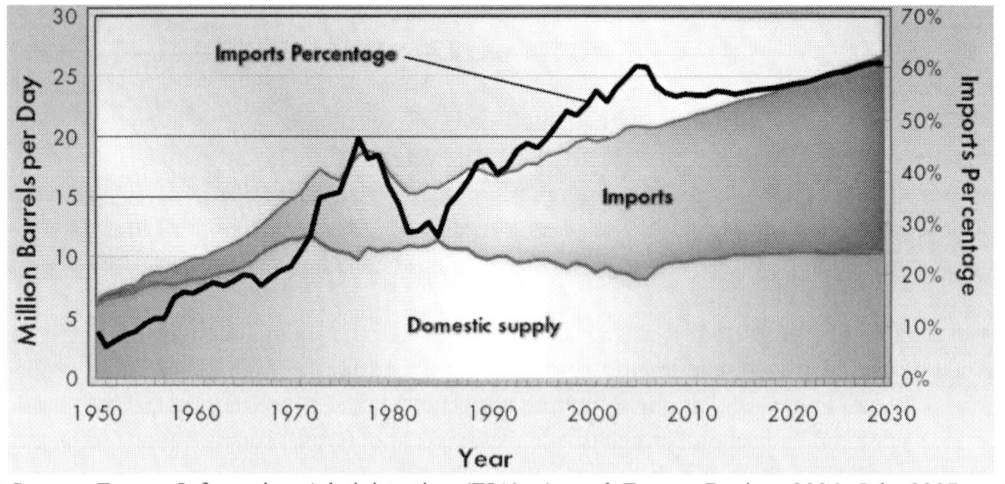

* Source: Energy Information Administration (EIA), Annual Energy Review 2006, July 2007, and Annual Energy Outlook 2007, February 2007.

Notes: Domestic supply includes crude oil production, lease condensate, natural gas plant liquids, refi nery process gain, ethanol, liquids produced from natural gas and coal, and other supplies such as other hydrocarbons, ethers, and biodiesel.

Figure 2. Domestic Liquid Fuels Supply and Imports — 1950–2030*.

DEPENDENCE ON FOREIGN OIL — A CONTINUING ANDGROWING CONCERN

America's economic well-being is heavily dependent upon the availability of secure and aff ordable transportation fuels such as gasoline, diesel fuel, and jet fuel. Th e United States is becoming increasingly reliant on imported oil, some of which comes from potentially unstable regions of the world, while at the same time our domestic crude oil production has decreased (Figure 1). Th ere also is growing global competition for petroleum as China and India continue their economic expansion. Finally, global energy delivery supply lines are getting longer, and exposure of these important lines to acts of terrorism will become more diffi cult to manage with time.

Acknowledging these factors, there is a growing consensus on the need to reduce U.S. dependence on imported oil, and to consider a portfolio approach of producing gasoline, diesel, and jet fuel from coal; utilizing oil shale; increasing domestic production of oil, gas, and biofuels; and increasing vehicle fuel economy.

The United States has an abundance of coal — approximately a 250-year supply at today's production rates. Figure 2 shows the relative magnitude of the domestic coal supplies in the United States compared to worldwide oil and coal reserves. Although the Middle East has the majority of proven oil reserves, the United States, Russia, China, India, and Australia control the largest coal reserves.

CTL Technology Basics

Approximately two barrels of clean diesel, gasoline, and jet fuel can be produced from a ton of coal. Th ere are three processes that can produce these fuels from coal:

- Indirect liquefaction, which breaks down the coal into simple molecules that are then combined to form liquid fuels;
- Direct liquefaction, which breaks down coal to the correct molecule size to form liquid fuels; and
- Hybrid concept, which incorporates technologies from both direct and indirect liquefaction processes.

Additionally, all three processes can effi ciently integrate carbon capture and storage technologies to mitigate global warming concerns.

In the indirect liquefaction process, coal fi rst is gasifi ed with oxygen and steam to produce synthesis gas — a mixture of carbon monoxide, hydrogen, and other compounds that is cleaned of impurities. The cleaned synthesis gas is sent to a water-gas shift reactor where the ratio of carbon monoxide-to-hydrogen is adjusted and optimized. Th e shift ed gas then is fed to the Fischer-Tropsch (F-T) reactors where the gas is converted to liquid fuels. The liquid fuels have a high cetane value (a measure of diesel fuel quality), and contain zero sulfur and essentially zero aromatic compounds. Th e process yields mostly diesel and jet fuels, which can be used in vehicles and airplanes, or blended into petroleum-derived diesel and jet fuels and subsequently used. Typically, carbon dioxide is captured during the water-gas shift and F-T reactions, making the process "carbon capture ready" and very amenable to carbon sequestration.

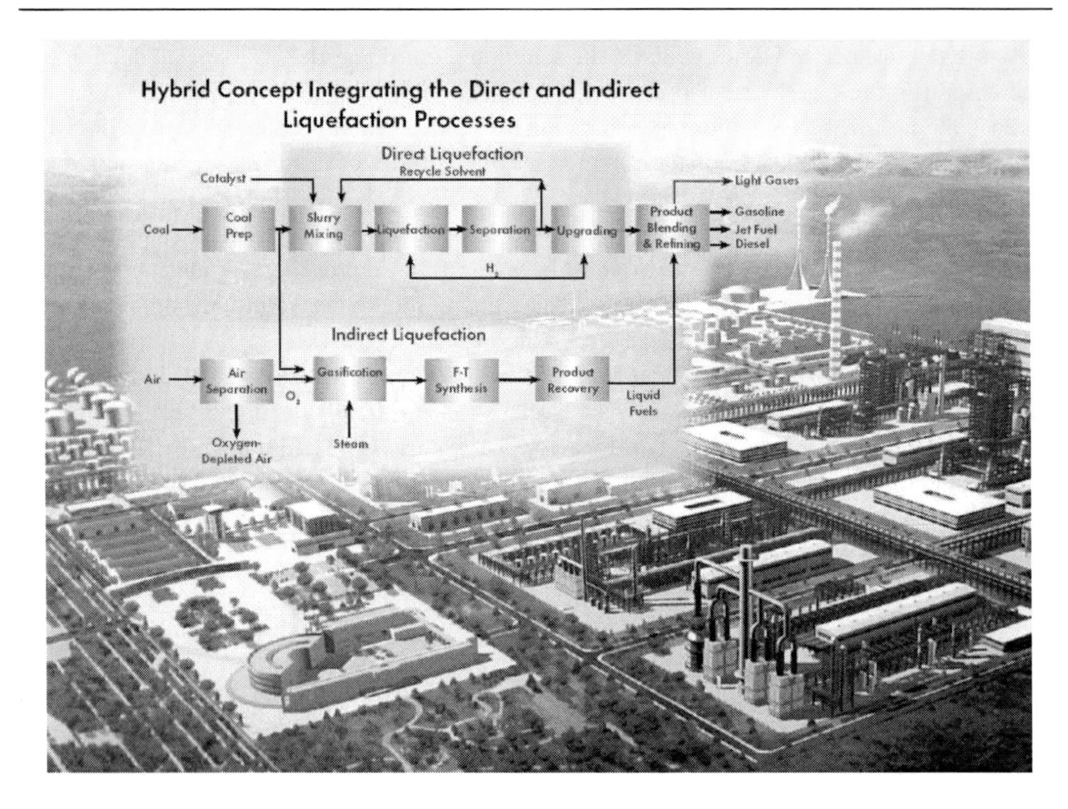

An alternative to the indirect liquefaction process is direct liquefaction, which converts coal at high temperature and pressure, in the presence of hydrogen and catalyst, to liquid fuels. Th is process results in more of, and a higher octane gasoline (measure of gasoline quality) compared to the indirect process. However, in order to meet current fuel quality requirements, some additional processing in a traditional oil refi nery may be required.

Finally, a hybrid process concept combines technologies from both the indirect and direct liquefaction processes.

Th ere are two key features of this hybrid process: 1) the hydrogen required for the direct process can be manufactured in the indirect process; and 2) the hybrid process yields both a high-quality diesel fuel from the indirect process and a highquality gasoline from the direct process.

Benefits of CTL Technology

- Reduce growing dependence on imported crude oil by using ample domestic coal reserves
- CTL technology has been demonstrated for more than 50 years
- CTL fuels can be zero-sulfur, zeroaromatic fuels, and can be used in existing engines
- CTL fuels can be distributed using the existing crude oil and product infrastructure
- Aside from liquid transportation fuels, CTL plants also can produce power and chemical feedstocks in "poly-generation" plants
- CTL plants and hydrogen from coal plants share many of the same technologies; therefore if a transition to a hydrogen economy occurs, investment in CTL plants will

not result in stranded investments since the plants can be converted to produce hydrogen

- Unique jet fuel qualities are of particular interest to the United States Air Force
- CTL fuels can be competitive with crude oil at $50/bbl
- Efficient integration of clean coal technologies and carbon capture and storage technology minimize environmental concerns

Artist Conception of the Shenhua Direct Coal Liquefaction Plant in China.

Focus on Environmental Issues and Greenhouse Gases

Coal liquefaction processes are designed to be responsive to environmental concerns, including global warming. The design of an indirect liquefaction plant provides a logical and costeff ective mechanism for carbon dioxide separation that can readily be used in a carbon capture and storage scheme. CTL technology also capitalizes on the decades of achievements of the DOE Clean Coal Program including gasifi cation technology advancements, oxygen production and separation, state-of-the-art effl uent gas treatment and emissions controls, novel hydrogen production technologies, and the latest carbon dioxide capture and storage developments.

CTL processes can also benefi t from the environmental advantages of co-fi ring biomass with coal, thus further minimizing carbon dioxide emissions. Since biomass resources such as switchgrass, hybrid poplar, and corn stover (the remaining parts of corn left in the fi eld such as the stalks and leaves) are considered renewable resources and produce no net CO_2 emissions, supplementing coal with some biomass allows the CTL facility to take advantage of the biomass CO_2 emissions benefit.

Evolution of CTL Technology

Coal liquefaction technology has its roots in Germany where direct liquefaction was developed by Fredrich Bergius in 1917, and indirect liquefaction was developed by Franz Fischer and Hans Tropsch in 1923. Th e process they developed is more commonly known today as the Fischer-Tropsch process.

CTL technology was originally utilized to assure fuel supplies during World War II. Th is process was costly and eventually abandoned aft er the war ended. The only other, and currently thriving, significant application of CTL technology is in South Africa, where more than 150,000 bpd of CTL fuels and chemicals are produced in vintage 1980s technology plants. The U.S. Department of Energy (DOE) had a successful CTL program in the 1980s coinciding with the spike in petroleum prices, but interest in the program subsided by the early 1990s as oil prices retreated. The DOE program during that time successfully developed several technologies to the demonstration-scale phase in partnership with industry. Until recently, high investment costs have limited further application of this technology.

Currently, China plans to make significant investments in CTL technology to enhance their energy security. Also, India is pursuing engineering studies for implementation of CTL technology. Both China and India have sizeable coal reserves, second only to the United States.

Technology Advancements

In the 1980s and 1990s, DOE research, development, and demonstration (RD&D) activities successfully developed several coal pilot scale liquefaction technologies and more recently, several companies have begun to initiate commercial-scale CTL activities in the United States and around the world. Of note, the People's Republic of China is aggressively pursuing CTL commercialization and is implementing technology developed under the DOE RD&D Program from the 1980s and 1990s.

In response to the concerns over climate change, there is growing emphasis on implementing carbon capture and storage technology and on co-feeding coal and biomass feedstocks to reduce the carbon footprint of CTL plants as noted below:

- DOE has an aggressive program on carbon sequestration which continues to make advancements in the safe, permanent, and secure storage of carbon dioxide from large central plants. As announced in October of 2007,
- DOE co-funded a $318 million new world class CO_2 sequestration program initiative aimed at demonstrating the technical viability of the CO_2 storage technologies and assuring environmental safeguards on an industrial scale.
- Co-feeding of coal and biomass to produce liquid fuels is a relatively new concept. DOE-sponsored R&D programs will play a key role in progressing this co-feeding concept in light of the limited data available on the processing of coal and biomass mixtures and technical and economic issues. Key issues include development of feedstock preparation and pre-treatment technologies needed for various coal and biomass feedstock types and feed mixture percentages of each and characterization of the solid, liquid, and gaseous products from the gasifier.

EXPANDED GOVERNMENT ROLE

Reliable and aff ordable energy is central to the United States' continued economic and national security. Th e role of the federal government is to help the nation meet its energy, scientific, environmental, and national security goals. Th is is achieved by developing and deploying new energy technologies to reduce dependence on foreign energy sources, in an environmentally responsive manner; ensuring U.S. competitiveness in the global marketplace; and encouraging entrepreneurship and innovation.

Expanded government eff orts in coal liquefaction can support this overall role, and are focused on four key areas: technology, operability, fi nancing, and education. Combined, these government eff orts should result in economically acceptable solutions that will provide important and environmentally sound alternatives for the nation's growing liquid fuel needs while maintaining U.S. technology leadership.

The United States Air Force has taken an active role in pursuing development and utilization of CTL fuels, having certified its B-52H Stratofortress aircraft on a 50/50 blend of JP-8 and F-T fuel. The Air Force plans to test and certify every airframe to operate on the blend by early 2011. Additionally, the Air Force has a goal of purchasing 50 percent of its fuel supply in 2016 from domestic synthetic fuel sources such as CTL. Another eff ort underway is a joint government-industry activity known as the Commercial Aviation Alternative Fuels Initiative (CAAFI). CAAFI participants include:

- Government: DOE, Department of Defense, Federal Aviation Administration, Department of Commerce, National Aeronautics and Space Administration
- Industry: Coal, biofuels, Air Transport Association, Aerospace Industries Association, Airports Council International-North America
- Universities and think tanks

CAAFI is pursuing alternative fuels for the purpose of securing a stable fuel supply, reducing environmental impacts, improving aircraft operations, and furthering research and analysis. CAAFI is structured into four panels — research, environmental, economics and business, and certifi cation and qualifi cation — which focus on achieving this purpose.

POSSIBLE FEDERAL GOVERNMENT ROLES

From a *technology* point of view: Supporting R&D to resolve the following issues regarding CTL technology: update process technology for latest gasifi cation developments, assure processing fl exibility to accommodate selected biomass feedstocks, conduct economic studies and analyses to determine most attractive process designs, and determine how best to integrate carbon capture and storage concepts.

From an *operability* point of view: Advance construction of fi rst-of-a-kind pioneer plants to demonstrate the overall technical feasibility and operability, and establish economic, technical, and environmental baselines to better defi ne the technology.

From a *fi nancing* point of view: Develop and implement the preferred fi nancial incentive package including investment credits, tax breaks, low cost fi nancing alternatives, and loan guarantees to reduce fi nancial risk and encourage industry investment.

From an *educational* point of view: Foster education and proactive communication programs to inform the public about CTL technologies, their products, and benefi ts, and address any concerns.

INDEX

2

A

B

C

D

E

F

G

H

I

J

K

L

M

N

O

P

T

U

V

W